MEMOIRS
of the
American Mathematical Society

ISSN 0065-9266 (print)
ISSN 1947-6221 (online)

Number 1557

q-Stability Conditions via q-Quadratic Differentials for Calabi-Yau-$\mathbb{X}$ Categories

Akishi Ikeda
Yu Qiu

April 2025 • Volume 308 • Number 1557 (fourth of 6 numbers)

AMS AMERICAN MATHEMATICAL SOCIETY

Library of Congress Cataloging-in-Publication Data

Cataloging-in-Publication Data has been applied for by the AMS.
See http://www.loc.gov/publish/cip/.

DOI: https://doi.org/10.1090/memo/1557

Memoirs of the American Mathematical Society

This journal is devoted entirely to research in pure and applied mathematics.

Subscription information. Beginning in 2024, Memoirs will be published monthly through 2026. Memoirs is also accessible from www.ams.org/journals. The 2025 subscription begins with volume 305 and consists of twelve mailings, each containing one or more numbers. Individual subscription prices for 2025 are as follows. For electronic only: US$1044. For paper delivery: US$1197. Add US$65 for delivery within the United States; US$200 for surface delivery outside the United States. Upon request, subscribers to paper delivery of this journal are also entitled to receive electronic delivery. For information on institutional pricing, please visit `https://www.ams.org/publications/journals/subscriberinfo`. Subscription renewals are subject to late fees. See `www.ams.org/journal-faq` for more journal subscription information. Each number may be ordered separately; please specify number when ordering an individual number.

Back number information. For back issues see `www.ams.org/backvols`.

Subscriptions and orders should be addressed to the American Mathematical Society, P. O. Box 845904, Boston, MA 02284-5904 USA. *All orders must be accompanied by payment.* Other correspondence should be addressed to 201 Charles Street, Providence, RI 02904-2213 USA.

Memoirs of the American Mathematical Society (ISSN 0065-9266 (print); 1947-6221 (online)) is published bimonthly (each volume consisting usually of more than one number) by the American Mathematical Society at 201 Charles Street, Providence, RI 02904-2213 USA. Periodicals postage paid at Providence, RI. Postmaster: Send address changes to Memoirs, American Mathematical Society, 201 Charles Street, Providence, RI 02904-2213 USA.

This publication is indexed in *Mathematical Reviews®*, *Zentralblatt MATH*, *Science Citation Index®*, *Science Citation Index™-Expanded*, *ISI Alerting Services℠*, *SciSearch®*, *Research Alert®*, *CompuMath Citation Index®*, *Current Contents®/Physical, Chemical & Earth Sciences.*

This publication is archived in *Portico* and *CLOCKSS*.

Printed in the United States of America.

∞ The paper used in this book is acid-free and falls within the guidelines established to ensure permanence and durability.

Visit the AMS home page at `https://www.ams.org/`

10 9 8 7 6 5 4 3 2 1 30 29 28 27 26 25

Contents

Abstract

Categorically, we introduce the Calabi-Yau-$\mathbb{X}$ categories $\mathcal{D}_{\mathbb{X}}$ of a graded marked surface $\mathbf{S}^{\lambda}$, as q-deformation of Haiden-Katzarkov-Kontsevich's topological Fukaya category $\mathcal{D}_{\infty}$ of $\mathbf{S}^{\lambda}$. We show that $\mathcal{D}_{\infty}$ can be identified with the cluster-$\mathbb{X}$ category associated to $\mathcal{D}_{\mathbb{X}}$.

Geometrically, we construct and identify the space of q-quadratic differentials on the logarithm surface $\log_{\mathbf{c}} \mathbf{S}^{\lambda}_{\Delta}$ with the space of induced q-stability conditions on $\mathcal{D}_{\mathbb{X}}$, for a complex parameter s satisfying $\mathrm{Re}(s) \gg 1$. When $s = N$ is an integer, the result gives an N-analogue of Bridgeland-Smith's result for realizing stability conditions on the Calabi-Yau-N orbit category $\mathcal{D}_{\mathbb{X}} \mathbin{/\!\!/} [\mathbb{X} - N]$ via CY-N type quadratic differentials.

When the genus of $\mathbf{S}$ is zero, the spaces of q-quadratic differentials can be also identified with framed Hurwitz spaces. As a byproduct, we confirms the conjectural almost Frobenius structure on spaces of q-stability conditions for type A.

Received by the editor January 30, 2022, and, in revised form, October 16, 2022.

Article electronically published on April 8, 2025.

DOI: https://doi.org/10.1090/memo/1557

2020 *Mathematics Subject Classification.* Primary 18G80, 30F30; Secondary 16E10, 16G20.

Key words and phrases. q-stability conditions, Calabi-Yau-$\mathbb{X}$ categories, cluster categories, q-quadratic differentials, twisted periods.

We would like to thank Tom Bridgeland, Yu-Wei Fan, Fabian Haiden, Bernhard Keller, Alastair King, Kyoji Saito, Atsushi Takahashi, and Yu Zhou for inspirational discussion.

This paper is supported by National Natural Science Foundation of China (Grant No. 12425104) and National Key R&D Program of China (No. 2020YFA0713000), Beijing Natural Science Foundation (Z180003), World Premier International Research Center Initiative (WPI initiative), MEXT, Japan and JSPS KAKENHI Grant Number JP16K17588.

CHAPTER 1

Introduction

In the prequel [**IQ**], we introduced q-stability conditions on a class of triangulated categories $\mathcal{D}_{\mathbb{X}}$ with a distinguished auto-equivalence $\mathbb{X}$.

In this paper, we study the case when $\mathcal{D}_{\mathbb{X}}$ are Calabi-Yau-$\mathbb{X}$ categories associated to graded decorated marked surfaces. We introduce q-quadratic differentials as counterpart of the q-stability conditions in prequel and prove they can be identified in the surface case.

1.1. Quadratic differentials as stability conditions

Motivated by Douglas' Π-stability of D-branes in string theory, Bridgeland [**B2**] introduced stability conditions on triangulated categories. Since then, the theory of stability conditions has played an important role in the study of mirror symmetry, Donaldson-Thomas invariants, cluster theory, etc. In particular, Kontsevich and Seidel observed that spaces of stability conditions and spaces of abelian/quadratic differentials share similar properties a couple of years ago (cf. [**BS**]). Recently, there are two seminal works in this direction:

- In [**BS**], Bridgeland-Smith (BS) showed that
$$\operatorname{Stab}^\circ \mathcal{D}_3(\mathbf{S}^3)/\operatorname{Aut} \cong \operatorname{Quad}_3(\mathbf{S}^3)$$
which is upgraded in [**KQ2**] as
$$\operatorname{Stab}^\circ \mathcal{D}_3(\mathbf{S}^3_\Delta) \cong \operatorname{FQuad}^\circ_3(\mathbf{S}^3_\Delta). \tag{1.1}$$
Here $\mathcal{D}_3(\mathbf{S}^3) \cong \mathcal{D}_3(\mathbf{S}^3_\Delta)$ is the Calabi-Yau-3 category associated to a (decorated) marked surface $\mathbf{S}^3$ (or $\mathbf{S}^3_\Delta$), $\operatorname{Stab}^\circ \mathcal{D}$ the space of stability conditions on a triangulated category $\mathcal{D}$, $\operatorname{Quad}_3(\mathbf{S}^3)$ the moduli space of (GMN/CY-3 type) quadratic differentials on $\mathbf{S}^3$ and $\operatorname{FQuad}_3(\mathbf{S}^3_\Delta)$ the $\mathbf{S}^3_\Delta$-framed version.
- In [**HKK**], Haiden-Katzarkov-Kontsevich (HKK) showed that
$$\operatorname{Stab}^\circ \mathcal{D}_\infty(\mathbf{S}^\lambda) \cong \operatorname{FQuad}_\infty(\mathbf{S}^\lambda), \tag{1.2}$$
where $\mathcal{D}_\infty(\mathbf{S}^\lambda)$ is the topological Fukaya category of a graded marked surface $\mathbf{S}^\lambda$, $\operatorname{Stab}^\circ \mathcal{D}_\infty(\mathbf{S}^\lambda)$ the space of stability conditions on $\mathcal{D}_\infty(\mathbf{S}^\lambda)$ and $\operatorname{Quad}_\infty(\mathbf{S}^\lambda)$ the moduli space of (exponential type) quadratic differentials on $\mathbf{S}^\lambda$.

These works share many similarities, which fit into the framework [**DHKK**] that relates dynamical systems and triangulated categories. For one thing, these categories are of topological Fukaya type. More precisely, the (spherical/indecomposable) objects in these categories correspond to (closed/graded) arcs on the surfaces (i.e., [**Q2**, Thm. 6.6], [**HKK**, Prop. 4.2]). Note that $\mathcal{D}_3(\mathbf{S}^3_\Delta)$ can be embedded into the derived Fukaya category of some symplectic manifold constructed from $\mathbf{S}^3_\Delta$ ([**S**], cf.

the survey [**Q4**]). For another, the constructions of a stability condition $\sigma = (Z, \mathcal{P})$ from a quadratic differential ξ are both roughly as follows:

- a saddle connection η with phase φ corresponds to (semi)stable objects X_η in the slicing $\mathcal{P}(\phi)$;
- the integral of the square root of ξ along (the double cover of) η gives the central charge of the corresponding stable object X_η:

$$Z(X_\eta) = \int_{\widetilde{\eta}} \sqrt{\xi}. \tag{1.3}$$

One of the main motivations of this paper is to give concrete links between these results [**BS**, **HKK**]. Categorically, we construct the (deformed) Calabi-Yau-$\mathbb{X}$ completion $\mathcal{D}_\mathbb{X}(\mathbf{S}^\lambda) := \mathcal{D}_\mathbb{X}(\mathbf{A})$ of $\mathcal{D}_\infty(\mathbf{S}^\lambda)$ via a full formal arc system $\mathbf{A}$ of the decorated version of $\mathbf{S}^\lambda$, such that $\mathcal{D}_3(\mathbf{S}^3_\Delta)$ is the (triangulated hull of the) orbit quotient:

$$\mathcal{D}_3(\mathbf{S}^3_\Delta) = \mathcal{D}_\mathbb{X}(\mathbf{S}^\lambda) \mathbin{/\!\!/} [\mathbb{X} - 3]. \tag{1.4}$$

Geometrically, we will q-deform HKK's result to produce a Calabi-Yau-s version of BS' result, that realizes q-stability conditions via q-quadratic differentials, for a complex parameter s.

A by-product we obtained along the way is that HKK's topological Fukaya category can be realized as the cluster category in the following sense:

$$\mathcal{D}_\infty(\mathbf{S}^\lambda) \cong \operatorname{per}(\mathbf{S}^\lambda) \cong \mathcal{C}_\mathbb{X}(\mathbf{S}^\lambda) := \operatorname{per}_\mathbb{X}(\mathbf{S}^\lambda)/\mathcal{D}_\mathbb{X}(\mathbf{S}^\lambda), \tag{1.5}$$

where $\operatorname{per}(\mathbf{S}^\lambda)$ is the Koszul dual of $\mathcal{D}_\infty(\mathbf{S}^\lambda)$ and $\mathcal{C}_\mathbb{X}(\mathbf{S}^\lambda)$ is the Calabi-Yau-$\mathbb{X}$ version of the cluster category, in the sense of Amiot-Guo-Keller [**A**, **K1**, **IY**]. This explains the cluster-like structure of $\operatorname{Stab}^\circ \mathcal{D}_\infty(\mathbf{S}^\lambda)$ in [**HKK**, § 6], cf. [**KQ1**, **Q1**].

1.2. The q-deformations

Given a triangulated category $\mathcal{D}_\infty$ with Grothendieck group $K(\mathcal{D}_\infty) \cong \mathbb{Z}^{\oplus n}$, Bridgeland shows that the set of all stability conditions (with support property) on $\mathcal{D}_\infty$ forms a complex manifold with dimension n. There is a local homeomorphism/coordinate

$$\begin{array}{rcl} \mathcal{Z}_\infty\colon \operatorname{Stab}\mathcal{D}_\infty & \longrightarrow & \operatorname{Hom}_\mathbb{Z}(K(\mathcal{D}_\infty), \mathbb{C}), \\ (Z, \mathcal{P}) & \mapsto & Z \end{array} \tag{1.6}$$

provided by the central charge Z of a stability condition $\sigma = (Z, \mathcal{P})$. To q-deform the notion of stability conditions and this result, we consider the Calabi-Yau-$\mathbb{X}$ version of $\mathcal{D}_\infty$, which is a triangulated category $\mathcal{D}_\mathbb{X}$ admitting a distinguish autoequivalence $\mathbb{X}$, that makes the Grothendieck group as

$$K(\mathcal{D}_\mathbb{X}) \cong R^{\oplus n}, \qquad R := \mathbb{Z}[q^{\pm 1}].$$

Recall that there are natural $\mathbb{C}$-action and Aut-action on the set of stability conditions.

A q-stability condition on $\mathcal{D}_\mathbb{X}$ consists of a stability condition σ and a complex number s satisfying the Gepner equation

$$\mathbb{X}(\sigma) = s \cdot \sigma, \tag{1.7}$$

(and a technical condition, the q-support property). In the prequel [**IQ**] we show that when fixing s, the q-stability conditions (σ, s) also form a complex manifold with dimension n with local homeomorphism/coordinate

$$\begin{array}{rcl} \mathcal{Z}_s\colon \operatorname{QStab}_s \mathcal{D}_{\mathbb{X}} & \longrightarrow & \operatorname{Hom}_R(K(\mathcal{D}_{\mathbb{X}}), \mathbb{C}_s), \\ (Z, \mathcal{P}, s) & \mapsto & Z. \end{array} \tag{1.8}$$

Here $\mathbb{C}_s$ is the complex plane equipped with the R-structure given by $q \cdot z := e^{\mathbf{i}\pi s} z$.

On the quadratic differential side, we also need to perform q-deformation. The moduli space $\operatorname{FQuad}_\infty(\mathbf{S}^\lambda)$ that corresponds to $\operatorname{Stab}\mathcal{D}_\infty(\mathbf{S}^\lambda)$ in (1.2) consists of quadratic differentials with exponential type singularities, locally of the form

$$e^{z^{-k}} z^{-l} g(z)\, \mathrm{d}\, z^2, \tag{1.9}$$

for some $(k, l) \in (\mathbb{Z}_{>0} \times \mathbb{Z})$ and some locally holomorphic non-vanishing function $g(z)$. The local homeomorphism/coordinate is given by the period map

$$\begin{array}{rcl} \Pi_\infty\colon \operatorname{FQuad}_\infty(\mathbf{S}^\lambda) & \longrightarrow & \operatorname{Hom}(\mathbf{H}_1(\mathbf{S}^\lambda, \partial\mathbf{S}^\lambda; \mathbb{Z}_{\mathrm{sp}}), \mathbb{C}), \\ \phi & \mapsto & \displaystyle\int \sqrt{\phi}, \end{array} \tag{1.10}$$

which corresponds to (1.6).

We would like to q-deform the singularity (1.9) as an s-pole of the form

$$e^{-k(s-2)-l}\, \mathrm{d}\, z^2 \tag{1.11}$$

together with k s-zeroes of the form $e^{(s-2)}\, \mathrm{d}\, z^2$. Analytically, we introduce a multi-valued quadratic differential Θ on $\mathbf{S}^\lambda$. Equivalently, we construct log-surface $\log_{\mathbf{c}} \mathbf{S}^\lambda_\Delta$ and view Θ as a single-valued quadratic differentials on $\log_{\mathbf{c}} \mathbf{S}^\lambda_\Delta$. The deck transformation, also denoted by q, on $\log_{\mathbf{c}} \mathbf{S}^\lambda_\Delta$ corresponds to the distinguish autoequivalence $\mathbb{X}$ and the equation (1.7) becomes

$$q^*(\Theta) = e^{\mathbf{i}\pi s} \cdot \Theta \tag{1.12}$$

in this setting. Then the local homeomorphism/coordinate is given by the period map

$$\begin{array}{rcl} \Pi_s\colon \operatorname{QQuad}_s(\log_{\mathbf{c}} \mathbf{S}^\lambda_\Delta) & \longrightarrow & \operatorname{Hom}_R(\widehat{\mathbf{H}}(\log_{\mathbf{c}} \mathbf{S}^\lambda_\Delta; \mathbb{Z}), \mathbb{C}_s), \\ \xi & \mapsto & \displaystyle\int \sqrt{\xi}, \end{array} \tag{1.13}$$

that corresponds to (1.8), where $\operatorname{QQuad}_s(\log_{\mathbf{c}} \mathbf{S}^\lambda_\Delta)$ is the moduli space of $\log_{\mathbf{c}} \mathbf{S}^\lambda_\Delta$-framed q-quadratic differentials.

When $\operatorname{Re}(s) \gg 1$, we use HKK's isomorphism (1.2) to prove the main result (Theorem 8.9) of the paper, which is a q-deformed generalization of BS' isomorphism (1.1).

THEOREM 1. *Suppose* $\operatorname{Re}(s) \gg 1$. *There is an isomorphism*

$$\operatorname{QQuad}_s(\log_{\mathbf{c}} \mathbf{S}^\lambda_\Delta) \xrightarrow{\cong} \operatorname{QStab}^\circ_s \mathcal{D}_{\mathbb{X}}(\mathbf{A})$$

between complex manifolds, where $\operatorname{QStab}^\circ_s \mathcal{D}_{\mathbb{X}}(\mathbf{A})$ *consists of (possibly many) connected components in the space* $\operatorname{QStab}_s \mathcal{D}_{\mathbb{X}}(\mathbf{A})$ *of q-stability conditions.*

1.3. Frobenius structures on genus zero Hurwitz spaces

A Hurwitz space HS is the moduli space of meromorphic functions on a Riemann surface S. In [**D1**, § 5], Dubrovin constructed Frobenius structures of Saito type (also called flat structure in [**Sa1**]) on Hurwitz spaces. The primitive form [**Sa1**, **SaTa**] (primary differential [**D1**]) on a Hurwitz space plays an essential role in his construction.

In Chapter 10, we identify the space of q-quadratic differentials with the regular locus of the Hurwitz space when the surface S is $\mathbb{P}^1$. As in Chapter 7, for q-quadratic differentials, we consider s-zeroes of the form $z^{(s-2)}\,\mathrm{d}\,z^2$ and s-poles of the form $z^{-k(s-2)-l}\,\mathrm{d}\,z^2$, which have the numerical data/polar type $(k,l) \in \mathbb{Z}_{\geq 1} \times \mathbb{Z}$. On the one hand, the positive integer k determines the complex structure of Hurwitz spaces. On the other hand, the choice of the integer l specifies the choice of the primitive form. Thus our introduction of the new parameter s and the division of the pole order into $k(s-2)$ plus l are essential for studying the moduli spaces of multi-valued quadratic differentials. The key numerical calculation is to match the winding numbers (in Section 7.5), which shows that a combination of k s-zeroes and one s-pole is indeed the q-deformation of HKK's exponential type singularity (1.9), with the same parameters (k,l).

On the regular locus $\mathrm{HS}_{\mathrm{reg}}$ of a Hurwitz space HS, we can consider the almost Frobenius structure [**D2**] as the almost dual of the original Frobenius structure. Under the identification between Hurwits spaces and moduli spaces, the twisted periods of the almost Frobenius structure on $\mathrm{HS}_{\mathrm{reg}}$ can be identified with the periods of q-quadratic differentials. Combining with our main result that identifies the moduli space of q-quadratic differentials with the space of q-stability conditions, we obtain the correspondence between the Hurwitz space and the space of q-stability conditions in this case.

In particular, twisted periods of the almost Frobenius structure should be identified with the central charges of q-stability conditions. In fact, we present a conjectural description of the almost Frobenius structure on the space QStab° of q-stability conditions on a Calabi-Yau-$\mathbb{X}$ category in the case of type ADE through isomorphism between an QStab° and the universal covering of the regular locus of the Cartan subalgebra of the corresponding type. One application of our q-deformation of quadratic differentials/stability conditions is to prove this conjecture in the type A_n case. The observation here is that the Cartan subalgebra of the type A_n can be identified with the unfolding of a type A_n singularity, which is a special case of a Hurwitz space.

In all, our new constructions provide an approach to understand almost Frobenius structure on the spaces of q-stability conditions.

1.4. Content

All the modules and categories will be over $\mathbf{k}$, an algebraically closed field. Denote $\mathbb{Z}/m\mathbb{Z}$ by $\mathbb{Z}_m$ for a positive integer m.

The paper is organized as follows.

In Part 1, we establish categorical setups:

- In Chapter 2, we recall the notations and results from the prequel [**IQ**] on q-deformation of stability conditions.

- In Chapter 3, we describe topological Fukaya categories from graded marked surfaces, partially following [**HKK**, **LP**, **IQZ**].
- In Chapter 4, we recall Keller's deformed Calabi-Yau completion from [**K1**] in the Calabi-Yau-$\mathbb{X}$ setting.
- In Chapter 5, we introduce the Calabi-Yau-$\mathbb{X}$ categories associated to graded marked surfaces, as q-deformations of the topological Fukaya categories. We recall technical results for such categories from [**IQZ**].

In Part 2, we study the geometric side of the story:

- In Chapter 6, we recall the theory of quadratic differentials of GMN type and exponential type on Riemann surfaces from [**BS**] and [**HKK**], respectively.
- In Chapter 7, we introduce q-quadratic differentials as q-deformations of quadratic differentials of exponential type. The key result (Theorem 7.21) here is that the numerical data (mainly, the winding numbers) for these two type of differentials do match.
- In Chapter 8, we prove the main results of the paper (Theorem 8.4 and Theorem 8.9), that the induced q-stability conditions on Calabi-Yau-$\mathbb{X}$ categories from surfaces can be identified with q-quadratic differentials.

In Part 3, we give applications:

- In Chapter 9, we show (Theorem 9.7) that when specializing $s = N$ to be a sufficiently large integer, the stability conditions on Calabi-Yau-N orbit categories can be realized as CY-N type quadratic differentials.
- In Chapter 10, we show (Theorem 10.9) that when the genus of the surface is zero, then framed q-quadratic differential can be identified with framed Hurwitz covers. In particular, this provides a solution of gluing q-stability conditions along the s-direction (cf. Corollary 10.12). Furthermore, when restricted to the disk case (type A_n), the moduli space of framed Hurwitz covers can be identified with the universal cover of the regular part of the space of polynomials of degree $n+1$ (cf. Proposition 10.15).
- In Chapter 11, we describe a conjectural almost Frobenius structure on spaces of q-stability conditions for Dynkin quivers and confirm it (Theorem 11.15) for the type A case using the result above on Hurwitz covers.

TABLE 1. List of notations

R	$\mathbb{Z}[q, q^{-1}]$
$\mathcal{D}_?(-)$	a Calabi-Yau-? category for $? \in \mathbb{Z} \cup \{\mathbb{X}, \infty\}$
$\sigma = (Z, \mathcal{P})$	a stability conditions with central charge Z and slicing $\mathcal{P}$
$\mathbf{S} = (\mathbf{S}, \mathbf{M}, \mathbf{Y})$	a marked surface, i.e., surface with sets of open/closed marked points
$\mathbf{S}^\lambda$	a graded marked surface with grading λ
$\mathbf{S}_\Delta$	a decorated marked surface (DMS), with decoration Δ
$\mathbf{c}$	a cut, i.e., a pairing between the set $\mathbf{Y}$ and the set Δ
$\mathbf{S}_\Delta = (\mathbf{S}_\Delta, \mathbf{c}, \Lambda)$	a graded DMS with grading Λ (compatible with $\mathbf{S}^\lambda$)
$\log_\mathbf{c} \mathbf{S}_\Delta^\lambda$	the log surface associated to $\mathbf{S}_\Delta$, w.r.t. $\mathbf{c}$
$\mathbf{S}_\Delta^{(m)}$	m^{th} sheet of $\log_\mathbf{c} \mathbf{S}_\Delta^\lambda$
$\mathbf{C}$	a marked surface $\mathbf{S}$ of genus 0
$\mathbf{D}$	a marked disk
$\mathfrak{N}(?)$	the numerical data of ?
$(\mathbf{k}, \mathbf{l})$	the polar type for exponential type/q- quadratic differentials
$\mathbf{A} = \{\widetilde{\gamma}_i\}$	a full formal open arc system for $\mathbf{S}^\lambda$
$\mathbf{A}^* = \{\widetilde{\eta}_i\}$	the dual full formal closed arc system
$\widetilde{\mathrm{CA}}(?)$	the set of graded closed arcs on ?
$\widehat{\mathbf{H}}(?) = \mathbf{H}_1(?)^-$	the hat homology of ?
$\mathcal{D}_\infty(\mathbf{A})$	the topological Fukaya category of $\mathbf{S}^\lambda$
$\mathcal{D}_\mathbb{X}(\mathbf{A})$	the Calabi-Yau-$\mathbb{X}$ category of $\mathbf{S}_\Delta$
(S, ϕ, h)	a quadratic differential ϕ on a Riemann surface S with framing h
$\mathrm{Core}(\phi)$	the core of a quadratic differential
$\Theta = (\mathrm{S}, \xi, h; s)$	a $\log_\mathbf{c} \mathbf{S}_\Delta^\lambda$-framed q-quadratic differential
$\mathrm{FQuad}_\infty(\mathbf{S}^\lambda)$	the moduli space of framed quadratic differential on $\mathbf{S}^\lambda$
$\mathrm{QQuad}_s^?(\log_\mathbf{c} \mathbf{S}_\Delta^\lambda)$	the moduli space of $\log_\mathbf{c} \mathbf{S}_\Delta^\lambda$-framed q-quadratic differential
$\mathrm{QStab}_s^? \mathcal{D}_\mathbb{X}(\mathbf{A})$	the space of certain type q-stability conditions on $\mathcal{D}_\mathbb{X}(\mathbf{A})$
$\mathrm{HS}(g, \mathbf{k})$	the moduli space of Hurwitz cover of genus g and polar type $\mathbf{k}$
$\mathrm{HS}(\log \mathbf{C}_\Delta)$	the moduli space of $\log \mathbf{C}_\Delta$-framed Hurwitz covers
$\mathcal{M}_n$	the space of polynomials of degree $n+1$
$?_{\mathrm{reg}}$	the regular part of space ?
$\mathbf{S}^N, \mathbf{S}_\Delta^N$	CY-N marked surface and DMS
$\overline{\mathbb{C}} = \mathbb{P}^1$	Riemann sphere

Part 1

Categorical Part

CHAPTER 2

q-Deformation of stability conditions

In this chapter, we recall the q-deformation of stability conditions from the prequel [**IQ**].

2.1. Bridgeland stability conditions

First we recall the definition of Bridgeland stability conditions on triangulated categories from [**B2**]. We assume that for a triangulated category $\mathcal{D}$, its Grothendieck group $K(\mathcal{D})$ is free of finite rank in this ection, i.e., $K(\mathcal{D}) \cong \mathbb{Z}^{\oplus n}$ for some n.

Definition 2.1. Let $\mathcal{D}$ be a triangulated category. A *stability condition* $\sigma = (Z, \mathcal{P})$ on $\mathcal{D}$ consists of a group homomorphism $Z\colon K(\mathcal{D}) \to \mathbb{C}$ called the *central charge* and a family of full additive subcategories $\mathcal{P}(\varphi) \subset \mathcal{D}$ for $\varphi \in \mathbb{R}$ called the *slicing* satisfying the following conditions:

(a) if $0 \neq E \in \mathcal{P}(\varphi)$, then $Z(E) = m(E)\exp(\mathbf{i}\pi\varphi)$ for some $m(E) \in \mathbb{R}_{>0}$,
(b) for all $\varphi \in \mathbb{R}$, $\mathcal{P}(\varphi + 1) = \mathcal{P}(\varphi)[1]$,
(c) if $\varphi_1 > \varphi_2$ and $A_i \in \mathcal{P}(\varphi_i)$ $(i = 1, 2)$, then $\mathrm{Hom}_{\mathcal{D}}(A_1, A_2) = 0$,
(d) for $0 \neq E \in \mathcal{D}$, there is a finite sequence of real numbers

(2.1) $$\varphi_1 > \varphi_2 > \cdots > \varphi_m$$

and a collection of exact triangles (*Harder-Narasimhan filtration*)

(2.2)

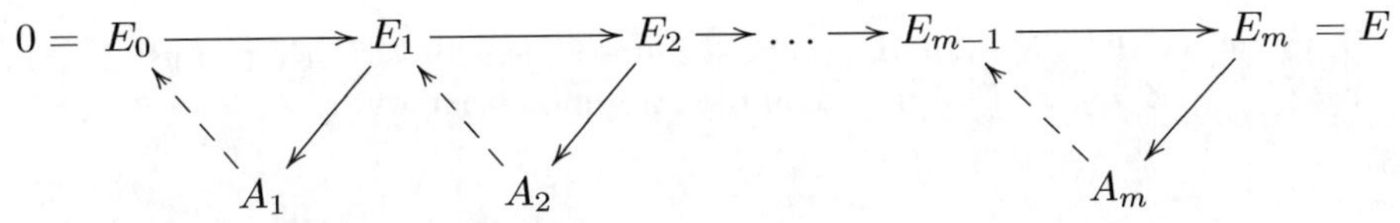

with $A_i \in \mathcal{P}(\varphi_i)$ for all i.

For each non-zero object $0 \neq E \in \mathcal{D}$, we define two real numbers by $\varphi_\sigma^+(E) := \varphi_1$ and $\varphi_\sigma^-(E) := \varphi_m$ where φ_1 and φ_m are determined by the axiom (d). Nonzero objects in $\mathcal{P}(\varphi)$ are called *semistable of phase* φ and simple objects in $\mathcal{P}(\varphi)$ are called *stable of phase* φ. For a non-zero object $E \in \mathcal{D}$ with extension factors $A_1, \ldots, A_m$ given by axiom (d), define the mass of E by

$$m_\sigma(E) := \sum_{i=1}^{m} |Z(A_i)|.$$

For a stability condition $\sigma = (Z, \mathcal{P})$, we introduce the set of semistable classes $\mathrm{ss}(\sigma) \subset K(\mathcal{D})$ by

(2.3)
$$\mathrm{ss}(\sigma) := \{\, \alpha \in K(\mathcal{D}) \,|\, \text{there is a semistable object } E \in \mathcal{D} \text{ such that } [E] = \alpha \,\}.$$

Let $\|\cdot\|$ be some norm on $K(\mathcal{D}) \otimes \mathbb{R}$. A stability condition $\sigma = (Z, \mathcal{P})$ satisfies the *support property* if there is some constant $C > 0$ such that

$$C \cdot |Z(\alpha)| > \|\alpha\| \tag{2.4}$$

for all $\alpha \in \mathrm{ss}(\sigma)$. We will assume that the stability conditions we considered always satisfy support property.

The key result by Bridgeland is the following.

THEOREM 2.2 ([**B2**], Theorem 1.2). *The projection map of taking central charges*

$$\mathcal{Z}\colon \operatorname{Stab}\mathcal{D} \longrightarrow \operatorname{Hom}(K(\mathcal{D}), \mathbb{C}), \qquad (Z, \mathcal{P}) \mapsto Z$$

is a local homeomorphism of topological spaces. In particular, $\mathcal{Z}$ induces a complex structure on $\operatorname{Stab}\mathcal{D}$.

There are two group actions on $\operatorname{Stab}\mathcal{D}$ commuting with each other. The first one is the natural $\mathbb{C}$ action

$$s \cdot (Z, \mathcal{P}) = (Z \cdot e^{-\mathbf{i}\pi s}, \mathcal{P}_{\mathrm{Re}(s)}),$$

where $\mathcal{P}_x(\phi) = \mathcal{P}(\phi + x)$. There is also a natural action on $\operatorname{Stab}\mathcal{D}$ induced by $\operatorname{Aut}\mathcal{D}$, namely:

$$\Phi(Z, \mathcal{P}) = \big(Z \circ \Phi^{-1}, \Phi(\mathcal{P})\big).$$

2.2. q-stability conditions

Let $\mathcal{D}_{\mathbb{X}}$ be a triangulated category with a distinguished auto-equivalence

$$\mathbb{X}\colon \mathcal{D}_{\mathbb{X}} \to \mathcal{D}_{\mathbb{X}}\,.$$

We will write $E[l\mathbb{X}]$ instead of $\mathbb{X}^l(E)$ for $l \in \mathbb{Z}$ and $E \in \mathcal{D}_{\mathbb{X}}$. Set $R = \mathbb{Z}[q^{\pm 1}]$ and define the action of R on $K(\mathcal{D}_{\mathbb{X}})$ by

$$q^n \cdot [E] := [E[n\mathbb{X}]].$$

Then $K(\mathcal{D}_{\mathbb{X}})$ has an R-module structure.

DEFINITION 2.3. A *q-stability condition* consists of a (Bridgeland) stability condition $\sigma = (Z, \mathcal{P})$ on $\mathcal{D}_{\mathbb{X}}$ and a complex number s satisfying

$$\mathbb{X}(\sigma) = s \cdot \sigma. \tag{2.5}$$

In the rest of this paper, we assume the following:

ASSUMPTION 2.4. *The Grothendieck group* $\mathrm{K}(\mathcal{D}_{\mathbb{X}})$ *is free of finite rank over R, i.e.,* $\mathrm{K}(\mathcal{D}_{\mathbb{X}}) \cong R^{\oplus n}$ *for some n.*

Similar to usual stability conditions, we also always require q-stability conditions satisfy the *q-support property* (see [**IQ**, Def. 3.8]).

Denote by $\operatorname{QStab}_s \mathcal{D}_{\mathbb{X}}$ the set of all q-stability conditions satisfying q-support property and with fixed s. Let

$$\operatorname{QStab}\mathcal{D}_{\mathbb{X}} = \bigcup_{s \in \mathbb{C}} \operatorname{QStab}_s \mathcal{D}_{\mathbb{X}}\,.$$

Let $\mathbb{C}_s$ denote the complex numbers with the R-module structure given by

$$q \cdot z := e^{\mathbf{i}\pi s} z$$

for $z \in \mathbb{C}_s$, that corresponds to the specialization (2.8). The following is our q-deformed version of Bridgeland's original result (Theorem 2.2).

THEOREM 2.5. [**IQ**, Thm. 3.10] *Let $\mathcal{D}_{\mathbb{X}}$ be a triangulated category with $\mathrm{K}(\mathcal{D}_{\mathbb{X}}) \cong R^{\oplus n}$ and fix $s \in \mathbb{C}$. The projection map of taking central charges*

$$\mathcal{Z}_s \colon \operatorname{QStab}_s \mathcal{D}_{\mathbb{X}} \longrightarrow \operatorname{Hom}_R(K(\mathcal{D}_{\mathbb{X}}), \mathbb{C}_s), \qquad (Z, \mathcal{P}) \mapsto Z$$

is a local homeomorphism of topological spaces. In particular, $\mathcal{Z}_s$ induces a complex structure on $\operatorname{QStab}_s \mathcal{D}_{\mathbb{X}}$.

2.3. $\mathbb{X}$-baric heart and q-stability conditions

DEFINITION 2.6 ($\mathbb{X}$-baric heart). An *$\mathbb{X}$-baric heart* $\mathcal{D}_\infty \subset \mathcal{D}_{\mathbb{X}}$ is a full triangulated subcategory of $\mathcal{D}_{\mathbb{X}}$ satisfying the following conditions:

(1) if $k_1 > k_2$ and $A_i \in \mathcal{D}_\infty[k_i\mathbb{X}]$ $(i = 1, 2)$, then $\operatorname{Hom}_{\mathcal{D}_{\mathbb{X}}}(A_1, A_2) = 0$,

(2) for $0 \neq E \in \mathcal{D}_{\mathbb{X}}$, there is a finite sequence of integers

$$k_1 > k_2 > \cdots > k_m$$

and a collection of exact triangles

$$\begin{array}{ccccccccc} 0 = E_0 & \longrightarrow & E_1 & \longrightarrow & E_2 \to \dots \to & E_{m-1} & \longrightarrow & E_m = E \\ & \nwarrow \swarrow & & \nwarrow \swarrow & & & \nwarrow \swarrow & \\ & A_1 & & A_2 & & & A_m & \end{array}$$

with $A_i \in \mathcal{D}_\infty[k_i\mathbb{X}]$ for all i.

Note that, by definition, the classes of objects in $\mathcal{D}_\infty$ span $K(\mathcal{D}_{\mathbb{X}})$ over R and there is a canonical isomorphism

$$K(\mathcal{D}_\infty) \otimes_{\mathbb{Z}} R \cong K(\mathcal{D}_{\mathbb{X}}). \tag{2.6}$$

Recall that the triangulated category $\mathcal{D}_{\mathbb{X}}$ is Calabi-Yau-$\mathbb{X}$ if $\mathbb{X}$ is the Serre functor (cf. (4.1)). For an $\mathbb{X}$-baric heart $\mathcal{D}_\infty$ in a Calabi-Yau-$\mathbb{X}$ category $\mathcal{D}_{\mathbb{X}}$, Condition (1) in Definition 2.6 is equivalent to

$$\operatorname{Hom}_{\mathcal{D}_{\mathbb{X}}}(A_1, A_2) = 0 \tag{2.7}$$

for $A_i \in \mathcal{D}_\infty[k_i\mathbb{X}]$ and $k_1 - k_2 \notin \{0, 1\}$.

For a fixed complex number $s \in \mathbb{C}$, consider the specialization

$$\begin{array}{rccc} q_s \colon & \mathbb{C}[q, q^{-1}] & \to & \mathbb{C}, \\ & q & \mapsto & e^{\mathbf{i}\pi s}. \end{array} \tag{2.8}$$

CONSTRUCTION 2.7. *Consider a triple $(\mathcal{D}_\infty, \widehat{\sigma}, s)$ consisting of an $\mathbb{X}$-baric heart $\mathcal{D}_\infty$, a (Bridgeland) stability condition $\widehat{\sigma} = (\widehat{Z}, \widehat{\mathcal{P}})$ on $\mathcal{D}_\infty$ and a complex number s. We construct*

1°. *the additive pre-stability condition $\sigma_\oplus = (Z, \mathcal{P}_\oplus)$ and*

2°. *the extension pre-stability condition $\sigma_* = (Z, \mathcal{P}_*)$,*

where

- *we first extend* $\widehat{Z}$ *to*

$$Z_q\colon = \widehat{Z} \otimes 1\colon K(\mathcal{D}_{\mathbb{X}}) \to \mathbb{C}[q, q^{-1}]$$

via (2.6) *and*

$$Z = q_s \circ (\widehat{Z} \otimes 1)\colon K(\mathcal{D}_{\mathbb{X}}) \to \mathbb{C}$$

gives a central charge function on $\mathcal{D}_{\mathbb{X}}$*;*
- *the pre-slicing* $\mathcal{P}_{\oplus}$ *is defined as*

$$\mathcal{P}_{\oplus}(\phi) = \operatorname{add}^s \widehat{\mathcal{P}}[\mathbb{Z}\mathbb{X}]\colon = \operatorname{add} \bigoplus_{k\in\mathbb{Z}} \widehat{\mathcal{P}}(\phi - k\operatorname{Re}(s))[k\mathbb{X}]; \tag{2.9}$$

- *the pre-slicing* $\mathcal{P}_*$ *is defined as*

$$\mathcal{P}_*(\phi) = \langle \widehat{\mathcal{P}}[\mathbb{Z}\mathbb{X}] \rangle^s\colon = \langle \widehat{\mathcal{P}}(\phi - k\operatorname{Re}(s))[k\mathbb{X}] \rangle. \tag{2.10}$$

Note that σ *does not necessarily satisfy condition (d) in Definition* 2.1 *and hence may not be a stability condition.*

Definition 2.8. **[IQ, Q5]** Given a slicing $\mathcal{P}$ on a triangulated category $\mathcal{D}$. Define the *global dimension* of $\mathcal{P}$ by

$$\operatorname{gldim} \mathcal{P} = \sup\{\phi_2 - \phi_1 \mid \operatorname{Hom}(\mathcal{P}(\phi_1), \mathcal{P}(\phi_2)) \neq 0\}. \tag{2.11}$$

For a stability condition $\sigma = (Z, \mathcal{P})$ on $\mathcal{D}$, its global dimension $\operatorname{gldim} \sigma$ is defined to be $\operatorname{gldim} \mathcal{P}$.

Definition 2.9. An *open q-stability condition* on $\mathcal{D}_{\mathbb{X}}$ is a pair (σ, s) consisting of a stability condition σ on $\mathcal{D}_{\mathbb{X}}$ and a complex parameter s such that

- $\sigma = \sigma_{\oplus}$ is an additive pre-stability condition induced from some triple $(\mathcal{D}_\infty, \widehat{\sigma}, s)$ as in Construction 2.7.

A *closed q-stability condition* on $\mathcal{D}_{\mathbb{X}}$ is a pair (σ, s) consisting of a stability condition σ on $\mathcal{D}_{\mathbb{X}}$ and a complex parameter s such that

- $\sigma = \sigma_*$ is an extension pre-stability condition induced from some triple $(\mathcal{D}_\infty, \widehat{\sigma}, s)$ as in Construction 2.7.

Theorem 2.10. **[IQ]** *Suppose that* $\mathcal{D}_{\mathbb{X}}$ *is Calabi-Yau-*$\mathbb{X}$ *with an* $\mathbb{X}$*-baric heart* $\mathcal{D}_\infty$*. A triple* $(\mathcal{D}_\infty, \widehat{\sigma}, s)$ *induces an open q-stability condition* (σ, s) *for* $\sigma = \sigma_{\oplus}$ *if and only if*

$$\operatorname{gldim} \widehat{\sigma} + 1 < \operatorname{Re}(s). \tag{2.12}$$

A triple $(\mathcal{D}_\infty, \widehat{\sigma}, s)$ *induces a closed q-stability condition* (σ, s) *for* $\sigma = \sigma_{\oplus}$ *if and only if*

$$\operatorname{gldim} \widehat{\sigma} + 1 \leq \operatorname{Re}(s) \tag{2.13}$$

We will still write

$$(\sigma, s) = (\mathcal{D}_\infty, \widehat{\sigma}, s) \otimes_{\oplus} R, \qquad (\sigma, s) = (\mathcal{D}_\infty, \widehat{\sigma}, s) \otimes_* R$$

for the two induced q-stability conditions in the above theorem if there causes no confusion. We will study subspaces $\operatorname{QStab}_s^{\oplus} \mathcal{D}_{\mathbb{X}}$ and $\operatorname{QStab}_s^{*} \mathcal{D}_{\mathbb{X}}$ of $\operatorname{QStab}_s \mathcal{D}_{\mathbb{X}}$ consisting of certain induced open and closed q-stability conditions.

CHAPTER 3

Topological Fukaya categories

3.1. Graded marked surface

We recall the relative notions and notations on graded marked surfaces, cf. [**IQZ, HKK, LP**]. Note that we will use a single marked point to denote a marked boundary arc in their setting, where if one performs the real blow-up (with respect to some quadratic differential) at the corresponding singularity, such a marked point corresponds to a line (isomorphic to $\mathbb{R}$).

DEFINITION 3.1. A *graded marked surface* $(\mathbf{S}, \mathbf{M}, \mathbf{Y}, \lambda)$ consists of the following data:

- $\mathbf{S}$ is a smooth oriented surface with $\partial\mathbf{S} \neq \emptyset$,
- $\mathbf{M}$ is a set of open marked points on $\partial\mathbf{S}$, such that $\mathbf{M} \cap \partial_i \neq \emptyset$ for each boundary component ∂_i of $\partial\mathbf{S}$ (cf. the blue bullet points in Figure 3.1).
- $\mathbf{Y}$ is a set of closed marked points on $\partial\mathbf{S}$, such that points in $\mathbf{M}$ and in $\mathbf{Y}$ are alternating in $\partial\mathbf{S}$ (cf. the red circle points in Figure 3.1). Let $\aleph := |\mathbf{Y}|$.
- λ is a grading (or line field) on $\mathbf{S}$, that is, a section $\lambda : \mathbf{S} \to \mathbb{P}T\mathbf{S}$ of the projectivized tangent bundle $\mathbb{P}T\mathbf{S}$.

For simplicity, we write $\mathbf{S}$ for the triple $(\mathbf{S}, \mathbf{M}, \mathbf{Y})$, called *marked surface*, and $\mathbf{S}^\lambda$ for the graded version. We will exclude the case when $\mathbf{S}$ is a disk with $\aleph = 2$.

A *morphism* $\mathbf{S}^\lambda \to \mathbf{S}'^{\lambda'}$ of graded marked surfaces is a pair $(f, \widetilde{f})$ consisting of an orientation preserving local diffeomorphism $f \colon \mathbf{S} \to \mathbf{S}'$, that preserves the sets of marked points setwise, and a (homotopy class of) path $\widetilde{f}$ from $f^*(\lambda')$ to λ in the space of sections of $\mathbb{P}T\mathbf{S}$. Every graded marked surface $\mathbf{S}$ admits an automorphism [1], known as the shift, given by the pair $(\mathrm{id}_{\mathbf{S}}, \overrightarrow{\pi})$, where restricted to the generator of $\pi_1(\mathbb{P}T_p\mathbf{S})$, $\overrightarrow{\pi}$ is rotating clockwise by an angle of π, for any $p \in \mathbf{S}$.

A *curve* in a graded marked surface $\mathbf{S}^\lambda$ is an immersion $c \colon I \to \mathbf{S}$ for a 1-manifold I. A *grading* $\widetilde{c}$ *on a curve* c is given by a family of (homotopy classes of) paths in $\mathbb{P}T_{c(t)}\mathbf{S}$ from $\lambda(c(t))$ to $\dot{c}(t)$, varying continuously with $t \in (0,1)$. The pair $(c, \widetilde{c})$ is called a *graded curve*, and will be simply denoted by $\widetilde{c}$ usually. The pushforward of a graded curve of a graded morphism $(f, \widetilde{f})$ is given by $(I, f \circ c, (c^*\widetilde{f}) \cdot \widetilde{c})$.

A point of transverse intersection of a pair $\big((I_1, c_1, \widetilde{c}_1), (I_2, c_2, \widetilde{c}_2)\big)$ of graded curves determines an integer as follows. Suppose $t_i \in I_i$ such that

$$c_1(t_1) = c_2(t_2) = p \in \mathbf{S} \quad \text{and} \quad \dot{c}_1(t_1) \neq \dot{c}_2(t_2) \in \mathbb{P}T_p\mathbf{S}. \tag{3.1}$$

We have the following homotopy classes of paths in $\mathbb{P}T_p\mathbf{S}$:

1°. $\widetilde{c}_1(t_1)$ from $\lambda(p)$ to $\dot{c}_1(t_1)$,
2°. $\widetilde{c}_2(t_2)$ from $\lambda(p)$ to $\dot{c}_2(t_2)$,

3°. κ from $\dot{c}_1(t_1)$ to $\dot{c}_2(t_2)$ given by counterclockwise rotation in $T_p\mathbf{S}$ by an angle less than π.

Define the *intersection index* of c_1, c_2 at p

$$\mathrm{ind}_p(c_1, c_2) = \widetilde{c}_1(t_1) \cdot \kappa \cdot \widetilde{c}_2(t_2)^{-1} \in \pi_1(\mathbb{P}T_p\mathbf{S}) \cong \mathbb{Z}. \tag{3.2}$$

It is well-defined if c_1 and c_2 are in general position.

3.2. Grading as cohomology/covering

The grading $\lambda : \mathbf{S} \to \mathbb{P}T\mathbf{S}$ can be interpreted as a cohomology class $[\lambda]$ in $\mathrm{H}^1(\mathbb{P}T\mathbf{S})$ in this case. The projection $\mathbb{P}T\mathbf{S} \to \mathbf{S}$ with $\mathbb{RP}^1 \backsimeq \mathrm{S}^1$-fiber gives a short exact sequence

$$0 \to \pi_1(\mathrm{S}^1) \to \pi_1(\mathbb{P}T\mathbf{S}) \to \pi_1(\mathbf{S}) \to 0,$$

or (taking the abelization)

$$0 \to \mathrm{H}_1(\mathrm{S}^1) \to \mathrm{H}_1(\mathbb{P}T\mathbf{S}) \to \mathrm{H}_1(\mathbf{S}) \to 0.$$

Dually, we have

$$0 \to \mathrm{H}^1(\mathbf{S}) \to \mathrm{H}^1(\mathbb{P}T\mathbf{S}) \xrightarrow{\pi_{\mathbf{S}}} \mathrm{H}^1(\mathrm{S}^1) = \mathbb{Z} \to 0. \tag{3.3}$$

[**LP2**, Lemma 1.2] shows that λ is determined by a class $[\lambda] \in \mathrm{H}^1(\mathbb{P}T\mathbf{S})$, induced from a split of $\pi_{\mathbf{S}}$ (i.e., $\pi_{\mathbf{S}}([\lambda]) = 1$ in $\mathrm{H}^1(\mathrm{S}^1)$). Such a data is equivalent to a split of $\mathrm{H}_1(\mathrm{S}^1) \to \mathrm{H}_1(\mathbb{P}T\mathbf{S})$ or a split of $\pi_1(\mathrm{S}^1) \to \pi_1(\mathbb{P}T\mathbf{S})$.

A more comprehensive way to think about λ is via the *Maslov ($\mathbb{Z}$-)covering*

$$\pi_\lambda \colon \mathbb{R}T\mathbf{S}^\lambda \to \mathbb{P}T\mathbf{S}$$

determined by λ, where $\mathbb{R}T\mathbf{S}^\lambda$ is the $\mathbb{R}$-bundle of $\mathbf{S}$ that can be constructed by gluing $\mathbb{Z}$ copies of $\mathbb{P}T\mathbf{S}$ cut out by λ. A morphism between two graded marked surfaces then consists of a map $f\colon \mathbf{S}^\lambda \to \mathbf{S}'^{\lambda'}$ such that it preserves the marked points and $[\lambda'] = f^*[\lambda']$, regarding $[\lambda] \in \mathrm{H}^1(\mathbb{P}T\mathbf{S})$, together with a map $\widetilde{f}$ between their Maslov covering that commutes with f. The automorphism/grading shift [1] is then the deck transformation of the Maslov covering.

In this description, a graded curve $\widetilde{c}$ is one of $\mathbb{Z}$ lifts in $\mathbb{R}T\mathbf{S}^\lambda$, of the tangent $\dot{c}$ of an usual curve c on $\mathbf{S}$. The intersection index $i = \mathrm{ind}_p(\widetilde{c}_1, \widetilde{c}_2)$ is the shift $[i]$ such that the lift $\widetilde{c}_2[i] \mid_p$ of $\widetilde{c}_2[i]$ at p is in the length one interval

$$(\widetilde{c}_1 \mid_p, \widetilde{c}_1[1] \mid_p) \subset \mathbb{R}T_p \cong \mathbb{R}.$$

3.3. Arc systems

We have the following notions.

- $\mathbf{M}$ (resp. $\mathbf{Y}$) divides $\partial\mathbf{S}$ into $\aleph$ *open (resp. closed) boundary arcs*, each of which contains a closed (resp. open) marked point.
- An *open (resp. closed) curve* γ in $\mathbf{S}$ is (the isotopy class of) a (non-trivial) curve connecting points in $\mathbf{M}$ (resp. in $\mathbf{Y}$), such that γ is in $\mathbf{S}^\circ = \mathbf{S} \setminus \partial\mathbf{S}$ except for its endpoints.
- An *(open/closed) arc* is an (open/closed) curve without self-intersection in $\mathbf{S}^\circ$.
- An open (resp. closed) *arc system* $\mathbf{A}$ is a collection of pairwise disjoint graded open (resp. closed) arcs.

- An open (resp. closed) *full formal* arc system is an open (resp. closed) arc system $\mathbf{A}$ such that it cuts $\mathbf{S}$ into polygons (called $\mathbf{A}$-polygons), each of which has exactly one open (resp. closed) boundary arc. See blue arcs in Figure 3.1.

The graded version of an open/closed curve a will usually be denoted by $\widetilde{a}$. Denote by $\mathrm{OA}(\mathbf{S}^\lambda)$ (resp. $\mathrm{CA}(\mathbf{S}^\lambda)$) the set of open (resp. closed) arcs on $\mathbf{S}^\lambda$, and by $\widetilde{\mathrm{OA}}(\mathbf{S}^\lambda)$ (resp. $\widetilde{\mathrm{CA}}(\mathbf{S}^\lambda)$) the set of graded ones.

Choose (and fix) an initial full formal open arc system $\mathbf{A} = \{\widetilde{\gamma}_j\}$. The dual graph of $\mathbf{A}$, denoted by $\mathbf{A}^* = \{\widetilde{\eta}_j\}$, is also a full formal closed arc system satisfying

$$\mathrm{Int}^\bullet(\widetilde{\gamma}_i, \widetilde{\eta}_j) = \mathrm{Int}^0(\widetilde{\gamma}_i, \widetilde{\eta}_j) = \delta_{ij},$$

where Int^d is the index-d intersection number and $\mathrm{Int}^\bullet = \sum_{d\in\mathbb{Z}} \mathrm{Int}^d$ is the total intersection number. As we only care about full formal arc systems, we will omit full formal for simplicity. The picture on the top of Figure 3.1 shows an example of *a dual arc system* $(\mathbf{A}, \mathbf{A}^*)$, where $\mathbf{A}$ consists of the blue arcs and $\mathbf{A}^*$ consists of red arcs.

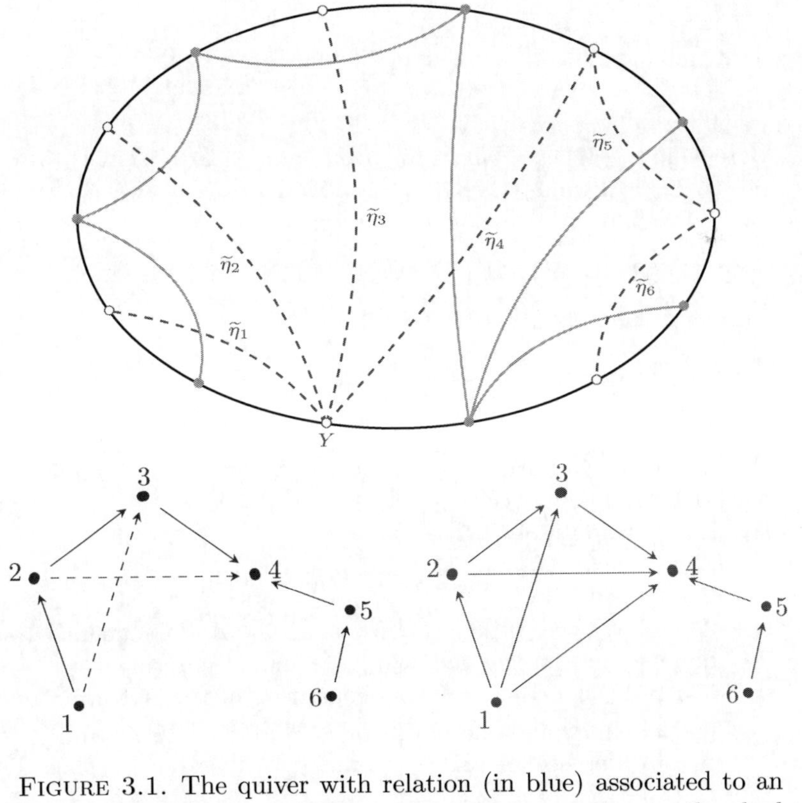

FIGURE 3.1. The quiver with relation (in blue) associated to an open arc system and the Ext-quiver corresponding to the dual closed arc system (in red)

3.4. Numerical data

DEFINITION 3.2. The *numerical data*

$$\mathfrak{N} = \mathfrak{N}(\mathbf{S}^\lambda) = (g, b; \mathbf{k}, \mathbf{l}; \mathrm{LP}_g)$$

of $\mathbf{S}^\lambda$ consists of

- the genus $g = g_\mathbf{S}$;
- the number $b = |\partial \mathbf{S}|$ of boundary components;
- the ordered multi-set $\mathbf{k} = (k_1, \ldots, k_b) \in \mathbb{Z}^b_{>0}$ of *orders*, where for each boundary component ∂_i, k_i is the number of marked points on ∂_i, satisfying

$$\sum_{i=1}^{b} k_i = |\mathbf{M}| = |\mathbf{Y}|; \tag{3.4}$$

- the ordered multi-set $\mathbf{l} = (l_1, \ldots, l_b) \in \mathbb{Z}^b$ of *indices*, where for each boundary component ∂_i, one can calculate l_i as follows: let $\{Y^i_j \mid j \in \mathbb{Z}_{k_i}\}$ be the closed marked points on ∂_i (in the clockwise order of the induced orientation of $\mathbf{S}$) and $\widetilde{\eta}^i_j = Y^i_j Y^i_{j+1}$ be the closed boundary arcs (with any grading), then

$$\begin{aligned} \mathrm{wind}_\lambda(\partial_i) = &\ 2 - l_i \\ = &\sum_{j \in \mathbb{Z}_{k_i}} \left(\mathrm{ind}_{Y^i_j}(\widetilde{\eta}^i_{j-1}, \widetilde{\eta}^i_j) - 1 \right). \end{aligned} \tag{3.5}$$

Here $\mathrm{wind}_\lambda(\partial_i)$ is the *(clockwise) winding number* of ∂_i with respect to λ (cf. [**LP2**, § 1.2] and [**HKK**, (3.21)]). More precisely, we have $\mathrm{wind}_\lambda(-) = \langle \lambda, - \rangle$, cf. (3.3), for the natural paring:

$$\langle -, - \rangle\colon \mathrm{H}^1(\mathbb{P}T\mathbf{S}) \times \mathrm{H}_1(\mathbb{P}T\mathbf{S}) \to \mathbb{Z}.$$

Note that the indices satisfy ([**LP2**, (1.5)])

$$\sum_{i=1}^{b} l_i = 4 - 4g. \tag{3.6}$$

We will use quadratic differentials to calculate the winding number $\mathrm{wind}_\lambda(\partial_i)$ and l_i in Section 7.5. Denote by $\mathbf{k} = (k_1, \ldots, k_b)$ and $\mathbf{l} = (l_1, \ldots, l_b)$ the data of (k_i, l_i).

- *the Lekili-Polishchuk (LP) data* LP_g, defined as follows:
  - if $g = 0$, then $\mathrm{LP}_0 = \emptyset$;
  - if $g = 1$, then $\mathrm{LP}_1 = \widetilde{A}$, where $\widetilde{A} := \widetilde{A}(\lambda)$ is defined to be the non-negative integer $\gcd\{\mathrm{wind}_\lambda(\gamma) \mid \text{non-separating } \gamma\}$;
  - if $g > 1$, then $\mathrm{LP}_g = (A, \kappa)$ consisting of the Arf invariant A (see [**LP2**, § 1.2]) and the indicator κ, where $\kappa = 0$ if $\mathrm{wind}^{(2)}_\lambda \equiv 0$ and $\kappa = 1$ otherwise. Here

    $$\mathrm{wind}^{(2)}_\lambda\colon \mathbf{H}_1(\mathbf{S}; \mathbb{Z}_2) \to \mathbb{Z}_2$$

    is the induced function ([**LP2**, Definition 1.2.1]).

EXAMPLE 3.3. Consider two genus zero cases.

- When $\mathbf{S}$ is a disk, which corresponds to a type A_n quiver, then the numerical data is

$$(g=0,\ b=1;\ \mathbf{k}=(n+1),\ \mathbf{l}=(4);\ \mathrm{LP}_0=\emptyset).$$

- When $\mathbf{S}$ is an annulus that corresponds to the affine type $\widetilde{A_{p,q}}$ quiver (with zero grading for all arrows), the numerical data is

$$(g=0,\ b=2;\ \mathbf{k}=(p,q),\ \mathbf{l}=(2,2);\ \mathrm{LP}_0=\emptyset).$$

In general, the indices can be chosen to be $\mathbf{l}=(2+w,2-w)$ for some $w\in\mathbb{Z}$ and the corresponding quiver $\widetilde{A_{p,q}}$ will be graded.

Note that in the case $g=0$, A and κ are irrelevant.

3.5. The topological Fukaya categories

The *graded quiver with relation* $(Q_{\mathbf{A}}^{(0)}, R_{\mathbf{A}}^{(0)})$ associated to an open arc system $\mathbf{A}$ is defined as follows:

- its vertices $\{i\}$ of $Q_{\mathbf{A}}^{(0)}$ correspond to the open arcs $\{\widetilde{\gamma}_i\}=\mathbf{A}$;
- its arrows correspond to the (anticlockwise) angle between arcs of $\mathbf{A}$ in the polygons of $\mathbf{S}$ cutting out by $\mathbf{A}$ and the grading of the arrows is given by the intersection index at the intersection (vertex of the angle);
- the quadratic relations are non-composable arrows=angles.

Let $\mathbf{k}$ be an algebraic closed field and $\mathcal{A}_{\mathbf{A}}^0=\mathbf{k}Q_{\mathbf{A}}^{(0)}/R_{\mathbf{A}}^{(0)}$. Next, we produce a differential graded algebra (we will write dga for short) from $\mathcal{A}_{\mathbf{A}}^0$, replacing the relation with differential.

Construction 3.4. [**Op**, Cons. 2.2] *Consider the basic finite dimensional* $\mathbf{k}$-*algebra* $\mathcal{A}_{\mathbf{A}}^0$. *Let* $Q_{\mathbf{A}}^{(1)}$ *be the (graded) quiver obtained from* $Q_{\mathbf{A}}^{(0)}$ *by adding the arrows corresponding to the (minimal set of) relations of* $R_{\mathbf{A}}^{(0)}$, *whose degrees are the degrees of the corresponding relations minus one, and a differential (of degree* 1*), such that* $\mathrm{H}^0(\mathcal{A}_{\mathbf{A}}^0)$ *is a quotient of* $\mathrm{H}^0(\mathbf{k}Q_{\mathbf{A}}^{(1)})$. *Now pick a generating set* $R_{\mathbf{A}}^{(1)}$ *for*

$$\ker\left(\mathrm{H}^1(\mathbf{k}Q_{\mathbf{A}}^{(1)})\to \mathrm{H}^0(\mathcal{A}_{\mathbf{A}}^0)\right)$$

to satisfy that $\mathbf{k}Q_{\mathbf{A}}^{(1)}/R_{\mathbf{A}}^{(1)}$ *is quasi-isomorphic to* $\mathcal{A}_{\mathbf{A}}^0$. *Iterating this process till we obtain a graded quiver* $Q_{\mathbf{A}}$ *(without relation) and a differential* d *such that* $\mathcal{A}_{\mathbf{A}}:=\mathbf{k}Q_{\mathbf{A}}$ *is quasi-isomorphic to* $\mathcal{A}_{\mathbf{A}}^0$.

Remark 3.5. By simple-projective duality

$$\mathrm{Irr}^d(P_i,P_j)\cong \mathrm{Ext}^{1-d}(S_j,S_i)^*,$$

$Q_{\mathbf{A}}$ can be also constructed from Ext-quiver ([**KQ1**, Def. 6.2]) of $\mathcal{A}_{\mathbf{A}}^0$. More precisely, Let $Q^{\mathrm{Ext}}(\mathcal{A}_{\mathbf{A}}^0)$ be the graded Ext-quiver of $\mathcal{A}_{\mathbf{A}}^0$, whose vertices are simple $\mathcal{A}_{\mathbf{A}}^0$ module and whose graded arrows are given by graded morphisms between them. Then $Q_{\mathbf{A}}$ is obtained from $Q^{\mathrm{Ext}}(\mathcal{A}_{\mathbf{A}}^0)$ by the degree changing: $d\mapsto 1-d$.

Moreover, $Q^{\mathrm{Ext}}(\mathcal{A}_{\mathbf{A}}^0)$ is the intersection quiver of $\mathbf{A}^*$, in the sense that its vertices one-one correspond to graded closed arcs in $\mathbf{A}^*$ and its graded arrows one-one correspond to graded (by index) intersections between them. For instance, the blue solid quiver in Figure 3.1 is $Q_{\mathbf{A}}^{(0)}$ while the relation $R_{\mathbf{A}}^{(0)}$ is presented by blue dashed arrows; the red quiver is the Ext-quiver of $Q_{\mathbf{A}}$.

Furthermore, such a duality topologically corresponds to the graph duality between $\mathbf{A}$ and $\mathbf{A}^*$ (see [**Q5**] for the Calabi-Yau-3 case of this correspondence).

Denote by $\mathrm{TFuk}(\mathbf{S}^\lambda)$ the *topological Fukaya category* of $\mathbf{S}^\lambda$, which in this case can be identified as

$$\mathrm{TFuk}(\mathbf{S}^\lambda) \cong \operatorname{per} \mathcal{A}_{\mathbf{A}} \cong \mathcal{D}_{fd}(\mathcal{A}_{\mathbf{A}}).$$

The above triangle equivalence follows from the fact that we require that any boundary component contains closed and open marked points (so that $\mathcal{A}_{\mathbf{A}}$ is homological smooth and proper). Lekili-Polishchuk proves the following.

THEOREM 3.6. [**LP2**] *Let $\mathbf{S}_i^{\lambda_i}$, $i = 1, 2$, be a graded marked surface with the numerical data $\mathfrak{N}(\mathbf{S}_i)$. Then $\mathfrak{N}(\mathbf{S}^{\lambda_1}) = \mathfrak{N}(\mathbf{S}^{\lambda_2})$ implies the triangle equivalence $\mathrm{TFuk}(\mathbf{S}_1^{\lambda_1}) \cong \mathrm{TFuk}(\mathbf{S}_2^{\lambda_1})$.*

We primarily interest in

$$\mathcal{D}_\infty(\mathbf{A}) \colon = \mathcal{D}_{fd}(\mathcal{A}_{\mathbf{A}}). \tag{3.7}$$

Denote by $\operatorname{Ind} \mathcal{C}$ the set of indecomposable objects in a category $\mathcal{C}$. The classification of objects in $\mathcal{D}_\infty(\mathbf{A})$ is given in [**HKK**]. However, we will need the version in [**IQZ**, § 6], which is slightly weaker on objects but with an additional intersection formula.

THEOREM 3.7. [**HKK,IQZ**] *There is an injection*

$$\widetilde{X}_\infty \colon \widetilde{\mathrm{CA}}(\mathbf{S}^\lambda) \longrightarrow \operatorname{Ind} \mathcal{D}_\infty(\mathbf{A}) \tag{3.8}$$

such that

$$\mathrm{Int}^d(\widetilde{\eta}_1, \widetilde{\eta}_2) = \dim \mathrm{Hom}^d(\widetilde{X}_\infty(\widetilde{\eta}_1), \widetilde{X}_\infty(\widetilde{\eta}_2)), \tag{3.9}$$

for any $\widetilde{X}_\infty(\widetilde{\eta}_i) \in \widetilde{\mathrm{CA}}(\mathbf{S}^\lambda)$.

Similarly, there is an injection

$$\widetilde{X}_\infty^\vee \colon \widetilde{\mathrm{OA}}(\mathbf{S}^\lambda) \longrightarrow \operatorname{Ind} \operatorname{per} \mathcal{A}_{\mathbf{A}},$$

but we will not use this map. Moreover, Theorem 3.7 implies that there is a canonical embedding (recall that we exclude the case when $\mathbf{S}$ is a disk with two closed marked points)

$$\mathrm{MCG}_\bullet(\mathbf{S}^\lambda) \to \mathrm{Aut}^\circ \mathcal{D}_\infty(\mathbf{A}).$$

Here, $\mathrm{MCG}_\bullet(\mathbf{S}^\lambda)$ is the *(marked) mapping class group* of $\mathbf{S}^\lambda$, i.e., the group of isotopy classes of morphisms of $\mathbf{S}^\lambda$, where all maps are required to fix $\mathbf{M}$ and $\mathbf{Y}$ setwise. And $\mathrm{Aut}^\circ \mathcal{D}_\infty(\mathbf{A})$ is the quotient group of the *auto-equivalence group* $\mathrm{Aut}\, \mathcal{D}_\infty(\mathbf{A})$ by those auto-equivalences of $\mathcal{D}_\infty(\mathbf{A})$ which acts trivially on the corresponding space of stability conditions $\mathrm{Stab}\, \mathcal{D}_\infty(\mathbf{A})$. Note that in this case, the auto-equivalence that preserves $\widetilde{X}_\infty(\widetilde{\mathrm{CA}}(\mathbf{S}^\lambda))$ will preserve $\mathrm{Stab}\, \mathcal{D}_\infty(\mathbf{A})$ since $\widetilde{X}_\infty(\widetilde{\mathrm{CA}}(\mathbf{S}^\lambda))$ contains stable objects (see Theorem 6.13 for more details).

CHAPTER 4

Deformed Calabi-Yau-$\mathbb{X}$ completion

In this section, we review Keller's deformed Calabi-Yau completion [**K1**].

4.1. $\mathbb{Z}^2$-graded differential modules and categories

We will work with *$\mathbb{Z}^2$-graded* **k**-*modules* ($\mathbb{Z}^2$-graded mods), So such a $\mathbb{Z}^2$-graded mod M admits a decomposition $M = \bigoplus_{(q,s)\in\mathbb{Z}^2} M_s^q$. Denote by $\deg m = (q,s) \in \mathbb{Z}^2$ the degree of an element $m \in M_s^q$. A $\mathbb{Z}^2$-graded morphism $f\colon M \to N$ between $\mathbb{Z}^2$-graded mods with degree (p,t) is a **k**-linear morphism such that $f(M_s^q) \subset N_{s+t}^{q+p}$. The first component of $\mathbb{Z}^2$-grading is the *cohomological grading* and denote the cohomological degree of $m \in M_s^q$ by $|m| = q$. Similarly $|f|$ denotes the cohomological degree of a $\mathbb{Z}^2$-graded morphism. The tensor product $M \otimes L$ is $(M\otimes L)_s^q :=$ $\bigoplus M_{s_1}^{q_1} \otimes L_{s_2}^{q_2}$ for $(q_1,s_1)+(q_2,s_2) = (q,s)$ and $f\otimes g\colon M\otimes L \to M'\otimes L'$ is

$$(f\otimes g)(m\otimes l)\colon = (-1)^{|m|\cdot|g|} f(m)\otimes g(l),$$

for $f\colon M\to M', g\colon L\to L'$, where only the cohomological degree effects the sign.

A *differential $\mathbb{Z}^2$-graded* **k**-*module* ($\mathbb{Z}^2$-dg-mod) is a pair consisting of a $\mathbb{Z}^2$-graded mod M and a (**k**-linear) *differential* map $\mathrm{d}_M\colon M\to M$ with degree $(1,0)$ such that $\mathrm{d}_M^2 = 0$. The cochain $\mathrm{Z}^q(M)$ and cohomology $\mathrm{H}^q(M)$ are defined with respect to the differential d (or the cohomological degree), which admits a $\mathbb{Z}$-grading (that corresponds the second grading on M)

$$\mathrm{Z}^q(M) = \bigoplus_s \mathrm{Z}_s^q(M), \quad \mathrm{H}^q(M) = \bigoplus_s \mathrm{H}_s^q(M).$$

REMARK 4.1. There are two natural shifts $M\{1\}$ and $M[1]$ on $\mathbb{Z}^2$-dg-mod M:

$$\begin{array}{ll} (M\{1\})_s^q = M_{s+1}^q, & \mathrm{d}_{M\{1\}} = \mathrm{d}_M \\ (M[1])_s^q = M_s^{q+1}, & \mathrm{d}_{M[1]} = -\mathrm{d}_M\,. \end{array}$$

A *$\mathbb{Z}^2$-dg-morphism* $f\colon M\to L$ between $\mathbb{Z}^2$-dg-mods is a **k**-linear map such that $\deg f = (0,0)$ and $f\circ \mathrm{d}_M = \mathrm{d}_L\circ f$, which induces a morphism (of graded modules) on cohomologies $\mathrm{H}^\bullet$. The tensor product $M\otimes L$ of two $\mathbb{Z}^2$-dg-mods is the tensor $\mathbb{Z}^2$-graded mod with $\mathrm{d}_{M\otimes L} = \mathrm{d}_M\otimes \mathrm{id}_L + \mathrm{id}_M\otimes \mathrm{d}_L$. The morphism space $\mathcal{H}om(M,L)$ is also a $\mathbb{Z}^2$-dg-mod, where the component $\mathcal{H}om_s^q(M,L)$ consists of all **k**-linear maps of degree (q,s) and

$$\mathrm{d}_f\colon = \mathrm{d}_M\circ f - (-1)^{|f|} f\circ \mathrm{d}_L\,.$$

A *$\mathbb{Z}^2$-graded differential category* ($\mathbb{Z}^2$-dg-cat) $\mathcal{A}$ is a **k**-category whose morphism spaces are $\mathbb{Z}^2$-dg-mods and whose compositions $\mathcal{A}(Y,Z)\otimes\mathcal{A}(X,Y)\to\mathcal{A}(X,Z)$ are $\mathbb{Z}^2$-dg-morphisms. A $\mathbb{Z}^2$-dg-functor $F\colon \mathcal{A}\to\mathcal{B}$ between $\mathbb{Z}^2$-dg-cats is given by a map between their objects and by morphisms of $\mathbb{Z}^2$-dg-mods $F(X,Y)\colon \mathcal{A}(X,Y)\to$ $\mathcal{B}(FX,FY)$ for $X,Y\in \operatorname{Obj}\mathcal{A}$.

The opposite $\mathbb{Z}^2$-dg-cat $\mathcal{A}^{op}$ of $\mathcal{A}$ has the same objects with morphisms $\mathcal{A}^{op}(X, Y) = \mathcal{A}(Y, X)$ and the composition of $f \in \mathcal{A}^{op}(Y, X)$ and $g \in \mathcal{A}^{op}(Z, Y)$ is given by $(-1)^{|f||g|}gf$. The cochain and homology category $\mathrm{Z}^0(\mathcal{A}), \mathrm{H}^0(\mathcal{A})$ of $\mathcal{A}$ have the same objects of $\mathcal{A}$ with morphisms $(\mathrm{Z}^0\,\mathcal{A})(X,Y) = \mathrm{Z}^0(\mathcal{A}(X,Y))$ and $(\mathrm{H}^0\,\mathcal{A})(X,Y) = \mathrm{H}^0(\mathcal{A}(X,Y))$ respectively.

For two $\mathbb{Z}^2$-dg-functors $F, G \colon \mathcal{A} \to \mathcal{B}$, the complex of graded morphisms $\mathcal{H}om(F,G)$ consists of a family of morphisms $\phi_X \colon \mathcal{B}(FX, FY)^n$ such that $(Gf)(\phi_X) = (\phi_Y)(Ff)$ with the induced differential from $\mathcal{B}(FX, FY)$. The set of morphisms between F and G is given by the set $\mathrm{Z}^0_0\,\mathcal{H}om(\mathcal{A},\mathcal{B})(F,G)$.

4.2. Derived categories of $\mathbb{Z}^2$-dg-cats

Denote by $\mathcal{C}_{dg}(A)$ the $\mathbb{Z}^2$-dg-cat of a $\mathbb{Z}$-graded $\mathbf{k}$-algebra $A = \bigoplus_{s\in\mathbb{Z}} A_s$. Let $\mathcal{A}$ be a small $\mathbb{Z}^2$-dg-cat. A (right) $\mathbb{Z}^2$-dg-$\mathcal{A}$-mod M is a $\mathbb{Z}^2$-dg-functor

$$M \colon \mathcal{A}^{op} \to \mathcal{C}_{dg}(\mathbf{k}).$$

There are also two natural shifts $[1], \{1\}$ for a $\mathbb{Z}^2$-dg-$\mathcal{A}$-mod M, the cohomological shift $[1]$ and the extra grading shift $\{1\}$, such that

$$M[1]\{1\}(X) := M(X)[1]\{1\} \in \mathcal{C}_{dg}(\mathbf{k}).$$

For each object X of $\mathcal{A}$, there is a right module $X^\wedge = \mathcal{A}(?, X)$ represented by X.

The $\mathbb{Z}^2$-dg-cat of $\mathcal{A}$ is defined to be

$$\mathcal{C}_{dg}(\mathcal{A}) \colon = \mathcal{H}om(\mathcal{A}^{op}, \mathcal{C}^{\mathbb{Z}}_{dg}(\mathbf{k}))$$

and we denote its morphism by $\mathcal{H}om_{\mathcal{A}}$. The category of $\mathbb{Z}^2$-dg-$\mathcal{A}$-mods $\mathcal{C}(\mathcal{A})$ has the $\mathbb{Z}^2$-dg-$\mathcal{A}$-mods as objects and the morphisms of $\mathbb{Z}^2$-dg-functors as morphisms and we have

$$\mathcal{C}(\mathcal{A}) = \mathrm{Z}^0\,\mathcal{C}_{dg}(\mathcal{A}).$$

The homotopy category of $\mathbb{Z}^2$-dg-$\mathcal{A}$-mods is $\mathcal{H}(\mathcal{A}) = \mathrm{H}^0\,\mathcal{C}_{dg}(\mathcal{A})$ whose morphisms are given by $\mathrm{H}^0_0\,\mathcal{H}om_{\mathcal{A}}$. Note that there are canonical isomorphisms $\mathcal{H}om(X^\wedge, M) \xrightarrow{\sim} M(X)$ and

$$\mathrm{H}(\mathcal{A})(X^\wedge, M[n]) \xrightarrow{\sim} \mathrm{H}^n\,M(X).$$

DEFINITION 4.2. Denote by $\mathcal{D}(\mathcal{A})$ the *derived category* of $\mathcal{C}(\mathcal{A})$ (or $\mathcal{H}(\mathcal{A})$) with respect to the class of quasi-isomorphisms and its homomorphisms by $\mathrm{Hom}_{\mathcal{D}(\mathcal{A})}$. An important fact is that $\mathcal{D}(\mathcal{A})$ also admits two equivalences, the triangulated shift $[1]$ and the extra grading shift $\{1\}$. We will write $[\mathbb{X}]$ for $\{1\}$ and $[m+l\mathbb{X}]$ for $[m]\circ[\mathbb{X}]^l$, where $m, l \in \mathbb{Z}$.

Recall that a $\mathbb{Z}^2$-dg-mod P is *cofibrant* if any morphism $P \to M$ factors through L for a surjective quasi-isomorphism $L \to M$; a $\mathbb{Z}^2$-dg-mod I is *fibrant* if any morphism $L \to I$ extends to M for an injective quasi-isomorphism $L \to M$.

LEMMA 4.3. [**K5**, Prop. 3.1] *For each $\mathbb{Z}^2$-dg-mod M, there is a quasi-isomorphism (known as cofibrant resolution) $\mathbf{p}M \to M$ with cofibrant $\mathbf{p}M$ and a quasi-isomorphism (known as fibrant resolution) $M \to \mathbf{i}M$ with fibrant $\mathbf{i}M$. Moreover, the projection functor $\mathcal{H}(\mathcal{A}) \to \mathcal{D}(\mathcal{A})$ admits a fully faithful left/right adjoint given by $M \mapsto \mathbf{p}M$ and $M \mapsto \mathbf{i}M$ respectively. Thus,*

$$\mathcal{H}(\mathcal{A})(\mathbf{p}L, M) = \mathrm{Hom}_{\mathcal{D}(\mathcal{A})}(L, M) = \mathcal{H}(\mathcal{A})(L, \mathbf{i}M).$$

4.3. The inverse dualizing complex and Calabi-Yau categories

Suppose that $\mathcal{A}$ is homologically smooth, i.e., $\mathcal{A}$ is perfect as an $\mathcal{A}^e$-mod. Here, $\mathcal{A}^e$ is the $\mathbb{Z}^2$-dg-cat $\mathcal{A} \otimes \mathcal{A}^{op}$, which admits an involution V^e such that

$$V^e(X,Y) = (Y,X) \quad \text{and} \quad V^e(f \otimes g) = (-1)^{|f|\cdot|g|} g \otimes f,$$

which is a preduality on $\mathcal{A}^e$.

A preduality $\mathbb{Z}^2$-dg-functor V on a $\mathbb{Z}^2$-dg-cat $\mathcal{A}$ is a $\mathbb{Z}^2$-dg-functor $V \colon \mathcal{A} \to \mathcal{A}^{op}$, such that the opposite functor V^{op} is a right adjoint of V. For a (right) $\mathbb{Z}^2$-dg-mod M, its dual left $\mathbb{Z}^2$-dg-mod $M^* \colon \mathcal{A} \to \mathcal{C}_{dg}(\mathbf{k})$ sends X to $\mathcal{H}om_{\mathcal{A}}(M, X^\wedge)$. The V-dual of M is a $\mathbb{Z}^2$-dg-mod

$$\begin{array}{rcl} M^\vee = M^* \circ V^{op} \colon \mathcal{A} & \to & \mathcal{C}_{dg}(\mathbf{k}), \\ X & \mapsto & \mathcal{H}om_{\mathcal{A}}(M, (V^{op}X)^\wedge). \end{array}$$

This induces a preduality functor, still denoted by V:

$$V \colon \mathcal{C}_{dg}(\mathcal{A}) \to \mathcal{C}_{dg}(\mathcal{A}^{op}).$$

Now let $\mathcal{A}$ be a small $\mathbb{Z}^2$-dg-cat. As $\mathbf{k}$ is a field, $\mathcal{A}$ is cofibrant over $\mathbf{k}$.

DEFINITION 4.4. The *inverse dualizing complex* $\Theta_{\mathcal{A}}$ is any cofibrant replacement of the image of $\mathcal{A}$ (considered as a right $\mathbb{Z}^2$-dg-$\mathcal{A}^e$-mod) under the total derived functor of V^e.

Denote by $\mathcal{D}_{fd}(\mathcal{A})$ the finite dimensional derived category of $\mathcal{A}$, which is the full subcategory of $\mathcal{D}(\mathcal{A})$ formed by the $\mathbb{Z}^2$-dg-mod M such that $\sum_i \dim \mathrm{H}^i(M)$ is finite.

LEMMA 4.5. [**K1**, Lemma 3.4] *Then for any* $L \in \mathcal{D}(\mathcal{A})$ *and* $M \in \mathcal{D}_{fd}(\mathcal{A})$, *there is a canonical isomorphism*

$$\mathrm{Hom}_{\mathcal{D}(\mathcal{A})}(L \otimes_{\mathcal{A}} \Theta_{\mathcal{A}}, M) \xrightarrow{\sim} \mathrm{D}\,\mathrm{Hom}_{\mathcal{D}(\mathcal{A})}(M, L),$$

where $\mathrm{D} = \mathrm{Hom}_{\mathbf{k}}(?, \mathbf{k})$.

Let $\mathcal{N} = m + l\mathbb{X}$. A triangulated category $\mathcal{D}$ is called *Calabi-Yau-$\mathcal{N}$* (CY-$\mathcal{N}$) if, for any objects X, Y in $\mathcal{D}$ we have a natural isomorphism

$$\mathfrak{S} : \mathrm{Hom}(X,Y) \xrightarrow{\sim} \mathrm{D}\,\mathrm{Hom}(Y, X[\mathcal{N}]). \tag{4.1}$$

Further, an object S is $\mathcal{N}$*-spherical* if $\mathrm{Hom}^{\mathbb{Z}^2}(S,S) = \mathbf{k} \oplus \mathbf{k}[-\mathcal{N}]$ and (4.1) holds functorially for $X = S$ and Y in $\mathcal{D}$, where

$$\mathrm{Hom}^{\mathbb{Z}^2}(X,Y) \colon = \bigoplus_{m,l \in \mathbb{Z}} \mathrm{Hom}(X, Y[m + l\mathbb{X}]).$$

By Lemma 4.5, if $\Theta_{\mathcal{A}} \cong \mathcal{A}[-\mathcal{N}]$, then $\mathcal{D}_{fd}(\mathcal{A})$ is Calabi-Yau-$\mathcal{N}$. We are in particular interested in the case when $\mathcal{N} = \mathbb{X}$ or $\mathcal{N} = N \in \mathbb{Z}$. In these two cases, there is a *twist functor* $\Phi_S \in \mathrm{Aut}\,\mathcal{D}$ for each $\mathcal{N}$-spherical object S, defined by

$$\Phi_S(X) = \mathrm{Cone}\left(S \otimes \mathrm{Hom}^{\mathbb{Z}^2}(S,X) \to X\right) \tag{4.2}$$

with inverse

$$\Phi_S^{-1}(X) = \mathrm{Cone}\left(X \to S \otimes \mathrm{Hom}^{\mathbb{Z}^2}(X,S)^\vee\right)[-1]. \tag{4.3}$$

Note that the graded dual of a graded $\mathbf{k}$-vector space $V = \oplus_{m,l \in \mathbb{Z}} V_{m,l}[m + l\mathbb{X}]$ is

$$V^\vee = \bigoplus_{m,l \in \mathbb{Z}} \mathrm{D}V_{m,l}[-m - l\mathbb{X}].$$

4.4. Deformed Calabi-Yau completion

Let $\Theta_{\mathcal{A}}$ be the inverse dualizing complex of $\mathcal{A}$ and $\theta = \Theta_{\mathcal{A}}[\mathbb{X}-1]$. The Calabi-Yau-$\mathbb{X}$ completion of $\mathcal{A}$ is the tensor DG category

$$\Pi_{\mathbb{X}}(\mathcal{A}) = T_{\mathcal{A}}(\theta)\colon = \mathcal{A} \oplus \theta \oplus (\theta \otimes_{\mathcal{A}} \theta) \oplus \cdots .$$

Moreover, let c be an element of the Hochschild homology

$$\mathrm{HH}_{\mathbb{X}-2}(\mathcal{A}) = \mathrm{Tor}^{\mathcal{A}^e}_{\mathbb{X}-2}(\mathcal{A}, \mathcal{A}) \tag{4.4}$$

or a closed morphism of degree 1 form Θ to $\mathcal{A}$ (i.e., in $\mathrm{Hom}_{\mathcal{D}(\mathcal{A}^e)}(\theta, \mathcal{A}[1])$ (cf. [**K1**, § 5.1]).

DEFINITION 4.6. The deformation $\Pi_{\mathbb{X}}(\mathcal{A}, c)$ obtained from $\Pi_{\mathbb{X}}(\mathcal{A})$ by adding c to the differential is called a *deformed Calabi-Yau completion* of $\mathcal{A}$ with respect to c.

THEOREM 4.7. [**Y**, Thm. 3.17] *Suppose $\mathcal{A}$ is finitely cellular. Assume that the element $c \in \mathrm{HH}_{\mathbb{X}-2}(\mathcal{A})$ can be lifted to an element $\tilde{c}$ of the negative cyclic homology $\mathrm{HN}_{\mathbb{X}-2}(\mathcal{A})$. Then the deformed Calabi-Yau completion $\Pi_{\mathbb{X}}(\mathcal{A}, c)$ is homologically smooth and Calabi-Yau-$\mathbb{X}$.*

REMARK 4.8. Recently, Keller [**K4**] corrected some error in [**K1**]. As remarked in [**K4**], the assumption in the above theorem can be satisfied in the case of Ginzburg dga which we will deal with in Chapter 5.

Moreover, we prove a lemma, which is a slightly weaker Calabi-Yau-$\mathbb{X}$ version of [**K1**, Lemma 4.4 and Remark 5.3].

DEFINITION 4.9. [**KQ1**, Def. 7.2] A functor $\mathcal{L}\colon \mathcal{D}(\mathcal{A}) \to \mathcal{D}(\Pi_{\mathbb{X}}(\mathcal{A}, c))$ is a *Lagrangian(-$\mathbb{X}$) immersion* if

- for $L, M \in \mathcal{D}_{fd}(\mathcal{A})$, we have

$$\mathrm{RHom}_{\Pi}(\mathcal{L}(L), \mathcal{L}(M)) = \mathrm{RHom}_{\mathcal{A}}(L, M) \oplus D\,\mathrm{RHom}_{\mathcal{A}}(M, L)[-\mathbb{X}]. \tag{4.5}$$

In particular, it is fully faithful restricted to the finite dimensional derived categories, i.e.,

$$\mathrm{Hom}_{\mathcal{D}_{fd}(\Pi_{\mathbb{X}}(\mathcal{A},c))}(\mathcal{L}(L), \mathcal{L}(M)) = \mathrm{Hom}_{\mathcal{D}_{fd}(\mathcal{A})}(L, M) \tag{4.6}$$

for any $L, M \in \mathcal{D}_{fd}(\mathcal{A})$.

LEMMA 4.10. *The canonical projection (on the first component) $\Pi_{\mathbb{X}}(\mathcal{A}, c) \to \mathcal{A}$ induces a Lagrangian immersion*

$$\mathcal{L}\colon \mathcal{D}_{fd}(\mathcal{A}) \to \mathcal{D}_{fd}(\Pi_{\mathbb{X}}(\mathcal{A}, c)).$$

Moreover, the image of $\mathcal{L}$ is an $\mathbb{X}$-baric heart of $\Pi_{\mathbb{X}}(\mathcal{A}, c)$.

PROOF. By definition, (4.5) holds for any (double) shifts of simple $\mathcal{A}$-modules. Denote by $\mathrm{Sim}\,\mathcal{D}_{fd}(\mathcal{A})$ the set of all shifts of simple $\mathcal{A}$-modules. Note that any object M in $\mathcal{D}_{fd}(\mathcal{A})$ admits a simple filtration with factors in $\mathrm{Sim}\,\mathcal{D}_{fd}(\mathcal{A})$ (which is a refinement of the canonical filtration with respect to the heart $\mathcal{A}-\ \mathrm{mod}$ -the category of $\mathcal{A}$-modules). Thus, (4.5) follows by induction (on the numbers of factors of simple filtration of M and L).

For the second statement: note that any object in $\mathcal{D}_{fd}(\Pi_{\mathbb{X}}(\mathcal{A}, c))$ has a filtration with factors in

$$\{\mathrm{Sim}\,\mathcal{D}_{fd}(\mathcal{A})[m + l\mathbb{X}]\}_{m,l\in\mathbb{Z}}.$$

Moreover, $\mathcal{A}$ has non-negative $\mathbb{X}$-grading. Thus $\langle \mathrm{Sim}\,\mathcal{D}_{fd}(\mathcal{A})[m]\rangle_{m\in\mathbb{Z}} = \mathcal{D}_{fd}(\mathcal{A})$ is an $\mathbb{X}$-baric heart. □

CHAPTER 5

q-Deformation of topological Fukaya categories

Recall that we have a graded marked surface $\mathbf{S}^\lambda = (\mathbf{S}, \mathbf{M}, \mathbf{Y}, \lambda)$ with dual arc systems $(\mathbf{A}, \mathbf{A}^*)$.

5.1. Quivers with superpotential

DEFINITION 5.1. The *$\mathbb{Z}^2$-graded quiver with superpotential* $(\widetilde{Q}_\mathbf{A}, W_\mathbf{A})$ is defined as follows:

- the vertices in $\widetilde{Q}_\mathbf{A}$ are arcs in $\mathbf{A}$;
- for each arrow $a\colon \widetilde{\gamma} \to \widetilde{\gamma}'$ in $\widetilde{Q}_\mathbf{A}$, add a dual arrow $a^*\colon \widetilde{\gamma}' \to \widetilde{\gamma}$, where their degrees in $\widetilde{Q}_\mathbf{A}$ are

$$\begin{cases} \widetilde{\deg}\, a = \deg a, \\ \widetilde{\deg}\, a^* = (2 - \mathbb{X}) - \deg a; \end{cases}$$

- for each $\mathbf{A}$-polygon D, let its edges be $\widetilde{\gamma}_1, \ldots, \widetilde{\gamma}_m$ in clockwise order (and follows by the open boundary arc of D). Then there are graded arrows

$$\widetilde{\gamma}_i \underset{a_{ji}^*}{\overset{a_{ij}}{\rightleftarrows}} \widetilde{\gamma}_j \quad \text{for} \quad 1 \le i < j \le m-1.$$

  in $\widetilde{Q}_\mathbf{A}$ and let (composing from left to right)

$$W_D = \sum_{1 \le i<j<k \le m} a_{ij} a_{jk} a_{ik}^*; \tag{5.1}$$

- there is also a loop $\widetilde{\gamma}^*$ at each vertex $\widetilde{\gamma}$ of degree $1 - \mathbb{X}$ in $\widetilde{Q}_\mathbf{A}$;
- the superpotential $W_\mathbf{A}$ is the sum of W_D's for all $\mathbf{A}$-polygons D.

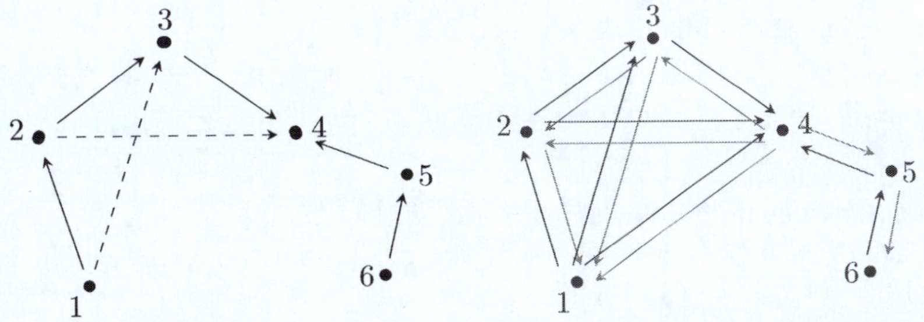

FIGURE 5.1. From the quiver with relation to the double graded quiver with potential

LEMMA 5.2. *The superpotential $W_{\mathbf{A}}$ defined as above is homogenous of degree $3-\mathbb{X}$.*

PROOF. It suffices to show that, for any $\mathbf{A}$-polygon D, W_D is homogenous of degree $3-\mathbb{X}$. Since $\mathbf{A}$ is full formal, there is a unique marked point Y_D in $\mathbf{Y}$ contained in D. Then the closed arcs in $\mathbf{A}^*$ starting at Y, denoted by $\widetilde{\eta}_1, \ldots, \widetilde{\eta}_m$ (in clockwise order, cf. Figure 3.1 for $m=4$), are the dual arcs of edges $\widetilde{\gamma}_1, \ldots, \widetilde{\gamma}_m$ of D. Note that an arc appears twice in $\{\widetilde{\eta}_i\}$ if it is a loop based at Y. Suppose that the indices of intersection at Y between graded arrows are

$$d_{ij} = \operatorname{ind}_Y(\widetilde{\eta}_i, \widetilde{\eta}_j), \quad \text{for} \quad 1 \le i < j \le m-1.$$

By the composition rule, we have

$$\deg d_{ij} = \sum_{k=i}^{j-1} d_{kk+1} \tag{5.2}$$

for any $1 \le i < j \le m$. Hence the corresponding arrow a_{ij} in $Q_{\mathbf{A}}$ has degree

$$\deg a_{ij} = 1 - d_{ij}, \tag{5.3}$$

and we have

$$\begin{aligned} & \widetilde{\deg}\, a_{ij} + \widetilde{\deg}\, a_{jk} + \widetilde{\deg}\, a_{ik}^* \\ = \; & (1-d_{ij}) + (1-d_{jk}) + \big((2-\mathbb{X}) - (1+d_{ik})\big) \\ = \; & 3 - \mathbb{X} \end{aligned}$$

for $1 \le i < j < k \le m$, as required. □

5.2. Calabi-Yau-$\mathbb{X}$ and cluster-$\mathbb{X}$ categories of surfaces

DEFINITION 5.3. The *Calabi-Yau-$\mathbb{X}$ Ginzburg $\mathbb{Z}^2$-dg algebra* $\Gamma_{\mathbf{A}}^{\mathbb{X}}$ is defined as follows:

- the underlying graded algebra of $\Gamma_{\mathbf{A}}^{\mathbb{X}}$ is the completion of the graded path algebra $\mathbf{k}\widetilde{Q}_{\mathbf{A}}$;
- the differential $\mathrm{d} = \mathrm{d}_{\mathbf{A}}$ of degree 1 is the unique continuous linear endomorphism satisfying $\mathrm{d}^2 = 0$ and the Leibniz rule (with respect to degree 1) and it takes the following values:
  - ▶ $\mathrm{d}\, a = \partial_{a^*} W_{\mathbf{A}}$ for $a \in (\widetilde{Q}_{\mathbf{A}})_1$;
  - ▶ $\mathrm{d} \displaystyle\sum_{\gamma \in (\widetilde{Q}_{\mathbf{A}})_0} \gamma^* = \sum_{a \in (\widetilde{Q}_{\mathbf{A}})_1} [a, a^*]$.

Note that we have the following convention:

- $\partial abc = (-1)^{\deg c} ab + (-1)^{\deg a} bc + (-1)^{\deg b} ca$ for any term abc in $W_{\mathbf{A}}$;
- $[a, a^*] = (-1)^{\deg a} aa^* + (-1)^{\deg a^*} a^* a$;

to ensure $\mathrm{d}^2 = 0$.

Let $\mathcal{D}(\Gamma_{\mathbf{A}}^{\mathbb{X}})$ be the derived category of $\mathbb{Z}^2$-dg $\Gamma_{\mathbf{A}}^{\mathbb{X}}$-modules, $\mathcal{D}_{fd}(\Gamma_{\mathbf{A}}^{\mathbb{X}})$ the finite dimensional derived category and $\operatorname{per} \Gamma_{\mathbf{A}}^{\mathbb{X}}$ the perfect derived category. They satisfy $\mathcal{D}_{fd}(\Gamma_{\mathbf{A}}^{\mathbb{X}}) \subset \operatorname{per} \Gamma_{\mathbf{A}}^{\mathbb{X}} \subset \mathcal{D}(\Gamma_{\mathbf{A}}^{\mathbb{X}})$.

Equivalently, we have the following. Let $\mathcal{R}_{\mathbf{A}}$ be the discrete $\mathbf{k}$-category associated to the vertex set of $Q_{\mathbf{A}}$ and $\mathcal{A}_{\mathbf{A}}$ the path category of $Q_{\mathbf{A}}$. Then $\Gamma_{\mathbf{A}}^{\mathbb{X}}$ is the tensor category over $\mathcal{R}_{\mathbf{A}}$ of the bimodule

$$\widetilde{Q}_{\mathbf{A}} = Q_{\mathbf{A}} \oplus (Q_{\mathbf{A}})^{\vee} [\mathbb{X}-2] \oplus \mathcal{R}_{\mathbf{A}}[\mathbb{X}-1].$$

Denote by $c_{\mathbf{A}}$ the image of $W_{\mathbf{A}}$ in (4.4).

THEOREM 5.4. [**K3**, Prop. 6.3] [**Y**, Thm. 3.17] *The Ginzburg dga* $\Gamma_{\mathbf{A}}^{\mathbb{X}}$ *is quasi-isomorphic to the deformed Calabi-Yau completion* $\Pi_{\mathbb{X}}(\mathcal{A}_{\mathbf{A}}, c_{\mathbf{A}})$. *Hence it is homologically smooth and Calabi-Yau-*$\mathbb{X}$.

Moreover, we obtain the following statement by applying Lemma 4.10.

COROLLARY 5.5. *The projection* $\Pi_{\mathbb{X}}(\mathcal{A}_{\mathbf{A}}, c) \to \mathcal{A}_{\mathbf{A}}$ *induces a Lagrangian immersion*

$$\mathcal{L}_{\mathbf{A}} \colon \mathcal{D}_{fd}(\mathcal{A}_{\mathbf{A}}) \to \mathcal{D}_{fd}(\Gamma_{\mathbf{A}}^{\mathbb{X}}) \tag{5.4}$$

and its image is an $\mathbb{X}$*-baric heart of*

$$\mathcal{D}_{\mathbb{X}}(\mathbf{A}) \colon = \mathcal{D}_{fd}(\Gamma_{\mathbf{A}}^{\mathbb{X}}). \tag{5.5}$$

By abuse of notations, we will identify $\mathcal{D}_{fd}(\mathcal{A}_{\mathbf{A}}) = \mathcal{D}_{\infty}(\mathbf{A})$ with its image under $\mathcal{L}_{\mathbf{A}}$ sometimes.

In the case for $\mathcal{D}_{\mathbb{X}} = \mathcal{D}_{\mathbb{X}}(\mathbf{A})$, we will study the principal part of the spaces of open/closed q-stability conditions.

DEFINITION 5.6. Denote by $\operatorname{QStab}_s^* \mathcal{D}_{\mathbb{X}}(\mathbf{A})$ the subspace of $\operatorname{QStab}_s \mathcal{D}_{\mathbb{X}}(\mathbf{A})$ consisting of closed q-stability conditions $\Psi(\sigma)$ for $\Psi \in \operatorname{ST} \mathcal{D}_{\mathbb{X}}(\mathbf{A})$ and σ is induced from some triple $(\mathcal{D}_{\infty}(\mathbf{A}), \widehat{\sigma}, s)$ as in Theorem 2.10 for some $\widehat{\sigma} \in \operatorname{Stab}^{\circ} \mathcal{D}_{\infty}(\mathbf{A})$ and $s \in \mathbb{C}$. Similarly, denote by $\operatorname{QStab}_s^{\oplus} \mathcal{D}_{\mathbb{X}}(\mathbf{A})$ the open version.

Applying the proof of [**IQ**, Thm. 6.7], we have the corresponding statement.

THEOREM 5.7. *There is a canonical triangle equivalence between the topological Fukaya category*

$$\operatorname{TFuk}(\mathbf{S}^{\lambda}) \colon = \operatorname{per} \mathcal{A}_{\mathbf{A}}$$

and the cluster-$\mathbb{X}$ *category*

$$\mathcal{C}(\Gamma_{\mathbf{A}}^{\mathbb{X}}) \colon = \operatorname{per} \Gamma_{\mathbf{A}}^{\mathbb{X}} / \mathcal{D}_{fd}(\Gamma_{\mathbf{A}}^{\mathbb{X}}). \tag{5.6}$$

5.3. Graded decorated marked surfaces

We fix the dual arc systems $(\mathbf{A}, \mathbf{A}^*)$ on $\mathbf{S}^{\lambda}$.

DEFINITION 5.8. We introduce the following notions.

- The *decorated marked surface* (DMS) $\mathbf{S}_{\Delta}$ of $\mathbf{S}$ is obtained from $\mathbf{S}$ by decorating a set $\Delta = \{Z_i\}_{i=1}^{\aleph}$ of points in the interior of $\mathbf{S}$, where $|\Delta| = |\mathbf{Y}| = |\mathbf{M}|$.
- A *(topological) cut* $\mathbf{c} = \{c_i\}$ is a set of curves on $\mathbf{S}$, pairing (connecting) points in Δ and $\mathbf{Y}$ with no intersections or self-intersections. See green arcs in Figure 5.2.
- A *grading* Λ on $\mathbf{S}_{\Delta}$ (with respect to $\mathbf{c}$) is a class in $\mathrm{H}^1(\mathbb{P}T(\mathbf{S} \setminus \Delta), \mathbb{Z}^2)$, with values
  - $(1, 0)$ on each (clockwise) loop $\{p\} \times \mathbb{RP}^1$ on $T_p(\mathbf{S} \setminus \Delta)$ for $p \notin \Delta$,
  - $(-2, 1)$ on each (clockwise) loop $l_Z \times \{x\}$ on $\mathbf{S}$ around any $Z \in \Delta, x \in \mathbb{RP}^1$, and
  - $(?, 0)$ on any simple closed curve α (i.e., a loop that is homeomorphic to S^1) on $\mathbf{S}$ that does not intersect with $\mathbf{c}$.

- A grading Λ on $\mathbf{S}_\Delta$ is compatible with (/induced from) the grading $\lambda \in \mathrm{H}^1(\mathbb{P}T\mathbf{S})$ if $\Lambda(\alpha) = (\lambda(\alpha), 0)$, for any simple closed curve α on $\mathbf{S}$ that does not intersect with $\mathbf{c}$.
- A graded DMS $\mathbf{S}_\Delta = (\mathbf{S}_\Delta, \mathbf{c}, \Lambda)$ associated to $\mathbf{S}^\lambda$ is a DMS such that $(\Lambda, \mathbf{c})$ are compatible with λ.

Open curves in $\mathbf{S}_\Delta$ still connect points in $\mathbf{M}$ while closed curves connect points in Δ. The notions of arcs and (full formal) open arc system are as before.

- Denote by $\mathrm{AC}(\mathbf{S}_\Delta)$ the set of admissible closed curves, where a closed curve is admissible if it does not cut out a once-decorated monogon by one of its self-intersections in $\mathbf{S}_\Delta - \Delta$.
- Denote by $\mathrm{CA}(\mathbf{S}_\Delta) \subset \mathrm{AC}(\mathbf{S}_\Delta)$ the set of closed arcs on $\mathbf{S}_\Delta$, which requires that they connect different decorations.
- The *mapping class group* $\mathrm{MCG}_\bullet(\mathbf{S}_\Delta)$ is the group of isotopy classes of diffeomorphisms of $\mathbf{S}_\Delta$, where all diffeomorphisms and isotopies are required to fix $\mathbf{M}$ and Δ setwise.
- The kernel of the forgetful map $\mathrm{MCG}_\bullet(\mathbf{S}_\Delta) \to \mathrm{MCG}_\bullet(\mathbf{S})$ is the *surface braid group* $\mathrm{SBr}(\mathbf{S}_\Delta)$, that is, the fundamental group of the configuration space of $|\Delta|$ points in (the interior of) $\mathbf{S}$, based at the set Δ.
- Each closed arc $\eta \in \mathrm{CA}(\mathbf{S}_\Delta)$ induces a (positive) *braid twist* $B_\eta \in \mathrm{MCG}_\bullet(\mathbf{S}_\Delta)$. The braid twist group $\mathrm{BT}(\mathbf{S}_\Delta)$ is the subgroup of $\mathrm{MCG}_\bullet(\mathbf{S}_\Delta)$ generated by $\{B_\eta \mid \eta \in \mathrm{CA}(\mathbf{S}_\Delta)\}$. Then $\mathrm{BT}(\mathbf{S}_\Delta) \subset \mathrm{SBr}(\mathbf{S}_\Delta)$, cf. [**Q2**].

We have the following characterization of cuts.

LEMMA 5.9. *The set* $\mathrm{Cut}(\mathbf{S}_\Delta)$ *of cuts is a* $\mathrm{SBr}(\mathbf{S}_\Delta)$*-torsor.*

PROOF. First, suppose that there is an element $b \in \mathrm{SBr}(\mathbf{S}_\Delta)$ and a cut $\mathbf{c}$ such that $b(\mathbf{c}) = \mathbf{c}$. As $\mathrm{SBr}(\mathbf{S}_\Delta) = \mathrm{MCG}_\bullet(\mathbf{S}_\Delta)/\mathrm{MCG}_\bullet(\mathbf{S})$, we can assume that b acts as the identity on $\mathbf{S}_\Delta - \mathbf{c}$ and hence b is the identity on $\mathbf{S}_\Delta$. Second, for any two cuts $\mathbf{c}_1$ and $\mathbf{c}_2$, one can construct an element $b \in \mathrm{SBr}(\mathbf{S}_\Delta)$, satisfying $b(\mathbf{c}_1) = b(\mathbf{c}_2)$, as follows:

- Take a homeomorphism/diffeomorphism $b_1 \colon \mathbf{S}_\Delta \to \mathbf{S}_\Delta$ that acts as identity outside the neighbourhood of $\mathbf{c}_1 = \{c_i^1\}$ and pulls decorations $\Delta = \{Z_i\}$ along $\{c_i^1\}$ to the neighbourhood of $\mathbf{Y} = \{Y_i\}$;
- Then take another homeomorphism $b_{21} \colon \mathbf{S}_\Delta \to \mathbf{S}_\Delta$ that moves the decorations (around $\mathbf{Y}$) on the arcs $\mathbf{c}_2 = \{c_i^2\}$;
- Finally, take a homeomorphism $b_2 \colon \mathbf{S}_\Delta \to \mathbf{S}_\Delta$ that moves the decorations along $\{c_i^2\}$ back to Δ.

The composition $b_2 \circ b_{21} \circ b_1$ gives the required b. This completes the proof. □

5.4. Topological log surfaces

To understand the grading of arcs/curves on $\mathbf{S}_\Delta$, we introduce log surface. In Section 7.4, we will discuss an alternative construction of log surface. Such a surface can be viewed as the q-deformation of $\mathbf{S}_\Delta$.

DEFINITION 5.10. The *(topological) log surface* $\log_{\mathbf{c}} \mathbf{S}_\Delta^\lambda$ of $\mathbf{S}_\Delta$ with respect to $\mathbf{c}$ is constructed as follows:

- For every $m \in \mathbb{Z}$, take a copy of $\mathbf{S}_\Delta$, denoted by $\mathbf{S}_\Delta^{(m)}$ with the cut $\mathbf{c}^{(m)}$. Cut it along $c_i^{(m)} \in \mathbf{c}^{(m)}$ from the decorating point.

- Gluing $c_{i+}^{(m)}$ with c_{i-}^{m+1} for $m \in \mathbb{Z}$ and $c_i \in \mathbf{c}$, we obtain $\log_{\mathbf{c}} \mathbf{S}_\Delta^\lambda$.
- $\log_{\mathbf{c}} \mathbf{S}_\Delta^\lambda$ inherits a ($\mathbb{Z}$-)grading $\Lambda_1 \in \mathrm{H}^1(\mathbb{P}T(\log_{\mathbf{c}} \mathbf{S}_\Delta^\lambda \setminus \Delta), \mathbb{Z})$ from Λ, which is the projection of Λ to the first $\mathbb{Z}$-coordinate on each sheet.
- There is a deck transformation q such that $q(\mathbf{S}_\Delta^{(m)}) = \mathbf{S}_\Delta^{(m+1)}$, which is also known as the Adams grading of $\log_{\mathbf{c}} \mathbf{S}_\Delta^\lambda$.
- Denote by $\pi_\Delta\colon \log_{\mathbf{c}} \mathbf{S}_\Delta^\lambda \to \mathbf{S}_\Delta$ the covering map. For any (graded) curve on $\mathbf{S}_\Delta$ (and is in a minimal position with respect to $\mathbf{c}$), there are $\mathbb{Z}$ lifts on $\log_{\mathbf{c}} \mathbf{S}_\Delta^\lambda$ via the deck transformation q. Denote by $\mathrm{CA}(\log_{\mathbf{c}} \mathbf{S}_\Delta^\lambda)$ and $\widetilde{\mathrm{CA}}(\log_{\mathbf{c}} \mathbf{S}_\Delta^\lambda)$ the set of all lifts of $\mathrm{CA}(\mathbf{S}_\Delta)$ and $\widetilde{\mathrm{CA}}(\mathbf{S}_\Delta)$ respectively. Similarly for $\mathrm{AC}(\log_{\mathbf{c}} \mathbf{S}_\Delta^\lambda)$ and $\widetilde{\mathrm{AC}}(\log_{\mathbf{c}} \mathbf{S}_\Delta^\lambda)$.
- The grading of $\log_{\mathbf{c}} \mathbf{S}_\Delta^\lambda$ inherits from the first grading Λ_1 of Λ on $\mathbf{S}_\Delta$, which will be denoted by $\log \lambda$.

The mapping class group $\mathrm{MCG}_\bullet(\mathbf{S}_\Delta)$ acts canonically on $\log_{\mathbf{c}} \mathbf{S}_\Delta^\lambda$ via the action on each sheet. Denote by $\mathrm{MCG}^\circ(\log_{\mathbf{c}} \mathbf{S}_\Delta^\lambda)$ the mapping class group of $\log_{\mathbf{c}} \mathbf{S}_\Delta^\lambda$, which is generated by q and homeomorphisms induced from $\mathrm{MCG}(\mathbf{S}_\Delta)$. Clearly, q is in the center of $\mathrm{MCG}^\circ(\log_{\mathbf{c}} \mathbf{S}_\Delta^\lambda)$.

A cut $\mathbf{c}$ is compatible with $\mathbf{A}$ if they do not intersect. From now on, we will fix the log surface $\log_{\mathbf{c}} \mathbf{S}_\Delta^\lambda$ with respect to an initial cut $\mathbf{c}$ that is compatible with the initial arc system $\mathbf{A}$ of $\mathbf{S}^\lambda$. In fact, $\mathbf{c}$ is uniquely determined by $\mathbf{A}$. Moreover, the (open full formal) arc system $\mathbf{A}$ is then also an arc system of $\mathbf{S}_\Delta$.

The lifts of graded closed arcs in $\log_{\mathbf{c}} \mathbf{S}_\Delta^\lambda$ provide a topological model for $\mathcal{D}_{\mathbb{X}}(\mathbf{A})$. More precisely, we have the following.

THEOREM 5.11. [**IQZ**, Thm. A] *There is a map*

$$\widetilde{X}_{\mathbb{X}}\colon \widetilde{\mathrm{AC}}(\log_{\mathbf{c}} \mathbf{S}_\Delta^\lambda) \to \mathrm{Ind}\,\mathcal{D}_{\mathbb{X}}(\mathbf{A}). \tag{5.7}$$

Then restricted to $\widetilde{\mathrm{CA}}(\log_{\mathbf{c}} \mathbf{S}_\Delta^\lambda)$, $\widetilde{X}_{\mathbb{X}}$ gives a bijection between the set of graded closed arcs and the set of reachable spherical objects (denoted by $\mathrm{Sph}\,\mathcal{D}_{\mathbb{X}}(\mathbf{A})$). Let $\mathrm{ST}\,\mathcal{D}_{\mathbb{X}}(\mathbf{A})$ be the spherical twists group, the subgroup of $\mathrm{Aut}\,\mathcal{D}_{\mathbb{X}}(\mathbf{A})$ generated by spherical twist Φ_S of $S \in \mathrm{Sph}\,\mathcal{D}_{\mathbb{X}}(\mathbf{A})$. Then X induces an isomorphism

$$\iota_{\mathbf{A}}\colon \mathrm{BT}(\mathbf{S}_\Delta) \cong \mathrm{ST}\,\mathcal{D}_{\mathbb{X}}(\mathbf{A}), \tag{5.8}$$

sending the braid twist B_η of a closed arc η to the spherical twist $\Phi_{\widetilde{X}_{\mathbb{X}}(\eta)}$.

Furthermore, similar to [**BS**, Thm. 9.9] (cf. [**KQ2**, § 4.4]), we can consider a quotient group $\mathrm{Aut}^\circ\,\mathcal{D}_{\mathbb{X}}(\mathbf{A})$ of certain subgroup of $\mathrm{Aut}\,\mathcal{D}_{\mathbb{X}}(\mathbf{A})$, that fits in the short exact sequence

$$0 \to \mathrm{ST}\,\mathcal{D}_{\mathbb{X}}(\mathbf{A}) \to \mathrm{Aut}^\circ\,\mathcal{D}_{\mathbb{X}}(\mathbf{A}) \to \mathrm{MCG}(\mathbf{S}) \to 0.$$

REMARK 5.12. The numerical data $\mathfrak{N}(\log_{\mathbf{c}} \mathbf{S}_\Delta^\lambda)$ of $\log_{\mathbf{c}} \mathbf{S}_\Delta^\lambda$ is given by $\mathfrak{N}(\mathbf{S}^\lambda)$ in Definition 3.2.

5.5. Topological immersion

The DMS $\mathbf{S}_\Delta$ can be thought as obtained from $\mathbf{S}$ by pulling the closed marked points along $\mathbf{c}$ that become decorations, cf. Figure 5.2. This operation induces a map

$$\mathbf{im}_{\mathbf{c}}\colon \widetilde{\mathrm{CA}}(\mathbf{S}^\lambda) \to \widetilde{\mathrm{CA}}(\mathbf{S}_\Delta^{(0)}) \subset \widetilde{\mathrm{CA}}(\log_{\mathbf{c}} \mathbf{S}_\Delta^\lambda) \tag{5.9}$$

in [**IQZ**, (6.2)], cf. [**IQZ**, Fig. 24], that takes a graded closed arc in $\mathbf{S}^\lambda$ to a graded closed arc in the 0-sheet $\mathbf{S}_\Delta^{(0)}$ of $\log_{\mathbf{c}} \mathbf{S}_\Delta^\lambda$.

Moreover, we have the following topological realization of Lagrangian immersion.

THEOREM 5.13. [**IQZ**, Thm. 6.6] *The following commutative diagram commutes*

$$\begin{array}{ccc} \widetilde{\mathrm{CA}}(\mathbf{S}^\lambda) & \xrightarrow{\mathbf{im_c}} & \widetilde{\mathrm{CA}}(\log_{\mathbf{c}} \mathbf{S}_\Delta^\lambda) \\ \widetilde{X}_\infty \downarrow & & \downarrow \widetilde{X}_{\mathbb{X}} \\ \operatorname{Ind} \mathcal{D}_\infty(\mathbf{A}) & \xrightarrow{\mathcal{L}_{\mathbf{A}}} & \operatorname{Ind} \mathcal{D}_{\mathbb{X}}(\mathbf{A}). \end{array} \tag{5.10}$$

Also notice that the full formal closed arc system $\mathbf{A}^*$ on $\mathbf{S}^\lambda$, dual to $\mathbf{A}$, becomes the closed arc system

$$\mathbf{A}_\Delta^* \colon = \mathbf{im_c}(\mathbf{A}^*)$$

on $\mathbf{S}_\Delta$, still dual to $\mathbf{A}$ (regarded on $\mathbf{S}_\Delta$).

REMARK 5.14. Similar to Remark 3.5, graph duality between $\mathbf{A}_\Delta^*$ and $\mathbf{A}$ corresponds to the simple-projective duality. More precisely, $\widetilde{Q}_{\mathbf{A}}$ can be obtained from the Ext-quiver $Q^{\mathrm{Ext}}(\Gamma_{\mathbf{A}}^{\mathbb{X}})$ by the degree changing: $d \mapsto 1-d$. And $Q^{\mathrm{Ext}}(\Gamma_{\mathbf{A}}^{\mathbb{X}})$ can be identified with the intersection quiver of $\mathbf{A}_\Delta^*$.

Furthermore, the Ext-quiver $Q^{\mathrm{Ext}}(\Gamma_{\mathbf{A}}^{\mathbb{X}})$ is the Calabi-Yau-$\mathbb{X}$ double, in the sense of [**KQ1**, Def. 6.2], of the Ext-quiver $Q^{\mathrm{Ext}}(\mathcal{A}_{\mathbf{A}}^0)$.

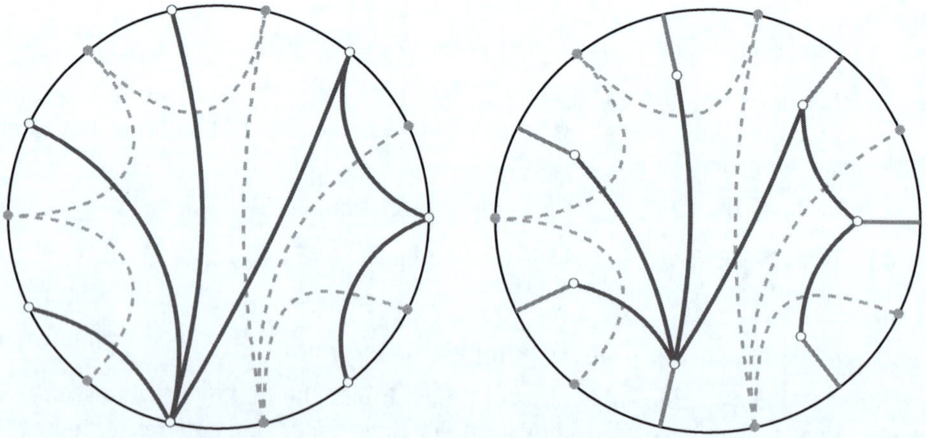

FIGURE 5.2. Constructing DMS $\mathbf{S}_\Delta$ via pulling closed marked points inside

Part 2

Geometric Part

CHAPTER 6

Quadratic differentials on Riemann surfaces

In this section, we review the theory of Bridgeland-Smith [**BS**] (cf. [**KQ2**]) and Haiden-Katzarkov-Kontsevich [**HKK**], that the spaces of stability conditions on Fukaya type categories associated to Riemann surfaces can be realized by the moduli spaces of framed quadratic differentials (of certain type) on the Riemann surfaces.

6.1. Preliminaries

Let S be a Riemann surface and ω_{S} its holomorphic cotangent bundle. A *meromorphic quadratic differential* ϕ on S is a meromorphic section of the line bundle ω_{S}^2. In terms of a local coordinate z on S, such a ϕ can be written as $\phi(z) = g(z)\,\mathrm{d}\,z^{\otimes 2}$, where $g(z)$ is a meromorphic function. Denote by $\mathrm{Zero}_j(\phi)$ the set of zeroes of ϕ of order j and $\mathrm{Zero}(\phi) = \bigsqcup_{j>0} \mathrm{Zero}_j(\phi)$. Similar for the set $\mathrm{Pol}(\phi) = \bigsqcup_{j>0} \mathrm{Pol}_j(\phi)$ of poles. Set $\mathrm{Crit}(\phi) := \mathrm{Zero}(\phi) \cup \mathrm{Pol}(\phi)$.

At a point of $\mathrm{S}^\circ = \mathrm{S} \setminus \mathrm{Crit}(\phi)$, there is a distinguished local coordinate ω, uniquely defined up to transformations of the form $\omega \mapsto \pm\omega + \mathrm{const}$, with respect to which one has $\phi(\omega) = \mathrm{d}\,\omega^{\otimes 2}$. In terms of a local coordinate z, we have $w = \int \sqrt{g(z)}\,\mathrm{d}\,z$. A quadratic differential ϕ on S determines the *ϕ-metric* on S°, defined locally by pulling back the Euclidean metric on $\mathbb{C}$ via the distinguished coordinate ω. Therefore, there are geodesics on S° and each geodesic has a constant phase, with respect to ω.

Next, we summarize the global structure of the horizontal trajectories and horizontal strip decompositions (cf. [**BS**, § 3] and [**HKK**, § 2]).

DEFINITION 6.1. A *horizontal trajectory* of ϕ on S° is a maximal horizontal geodesic $\gamma\colon (0,1) \to \mathrm{S}^\circ$ with respect to the ϕ-metric. The horizontal trajectories of a meromorphic quadratic differential ϕ provide the *horizontal foliation* on S.

The following are types of trajectories of a quadratic differential ϕ we are interested in:

- *saddle trajectories* with both ends in $\mathrm{Zero}(\phi)$;
- *separating trajectories* with one end in $\mathrm{Zero}(\phi)$ and the other in $\mathrm{Pol}(\phi)$;
- *generic trajectories* with both ends in $\mathrm{Pol}(\phi)$;
- *closed trajectories* that are simple closed curves;
- *recurrent trajectories* that are recurrent in at least one direction.

DEFINITION 6.2. By removing all horizontal separating trajectories (which are finitely many) from S°, the remaining open surface splits as a disjoint union of connected components. Each component is one of the following types:

- a *half-plane*, i.e., it is isomorphic to $\{z \in \mathbb{C} \colon \operatorname{Im}(z) > 0\}$ equipped with the differential $\mathrm{d}\, z^{\otimes 2}$. It is swept out by generic trajectories which connect a fixed pole to itself.
- a *horizontal strip*, i.e., it is isomorphic to $\{z \in \mathbb{C} \colon a < \operatorname{Im}(z) < b\}$ equipped with the differential $\mathrm{d}\, z^{\otimes 2}$. It is swept out by generic trajectories which connect two (not necessarily distinct) poles.
- a *ring domain*, i.e., it is isomorphic to $\{z \in \mathbb{C} \colon a < |z| < b\} \subset \mathbb{C}^*$ equipped with the differential $r\, \mathrm{d}\, z^{\otimes 2}/z^2$ for some $r \in \mathbb{R}_{<0}$. It is swept out by closed trajectories.
- a *spiral domain*, defined to be the interior of the closure of a recurrent trajectory.

We call this union the *horizontal strip decomposition* of S with respect to ϕ.

A quadratic differential ϕ on S is called *saddle-free*, if it has no saddle trajectory. Note the following:

- In each horizontal strip, the trajectories are isotopy to each other.
- If ϕ is saddle-free, then $\operatorname{Pol}(\phi)$ must be nonempty and, by [**BS**, Lemma 3.1], ϕ has no closed or recurrent trajectories. Thus, in the horizontal strip decomposition of S with respect to ϕ, there are only half-planes and horizontal strips.
- If ϕ is saddle-free, then the boundary of any strips consists of separating trajectories.
- If ϕ is saddle-free, then there is a unique geodesic in each horizontal strip, the *saddle connection*, connecting the two zeroes on its boundary.

6.2. Classical type of singularities in the Calabi-Yau-3 setting

DEFINITION 6.3. A *classical type of singularity* p of a meromorphic quadratic differential has the local form

$$z^k g(z)\, \mathrm{d}\, z^{\otimes 2}, \tag{6.1}$$

where $g(z)$ is some non-zero holomorphic function and

- if $k > 0$, then p is a *zero* of order k;
- if $k < 0$, then p is a *pole* of order k.

The foliation near a singularity p is shown in Figure 6.1, where p is a zero of order 1 and 2 in the left two pictures and p is a pole of order 3 and 4 in the right two pictures.

Next, we briefly review Bridgeland-Smith's result (cf. [**BS, KQ2**]). For simplicity, we will not consider the case when the pole has order 1 or 2.

DEFINITION 6.4. A *GMN differential* on S is a meromorphic differential with classical type of singularities satisfying:

- all the zeroes are simple zeroes (i.e., of order 1);
- the order of any pole is at least 3.

The *real (oriented) blow-up* of (S, ϕ) is a differentiable surface S^ϕ obtained from the underlying differentiable surface by replacing a pole $P \in \operatorname{Pol}(\phi)$ with order $\operatorname{ord}_\phi(P) > 3$ by a boundary ∂_P, where the points on the boundary correspond to the real tangent directions at P. We mark the points on ∂_P that correspond to the distinguished tangent directions. Thus there are $\operatorname{ord}_\phi(P) - 2$ (open) marked

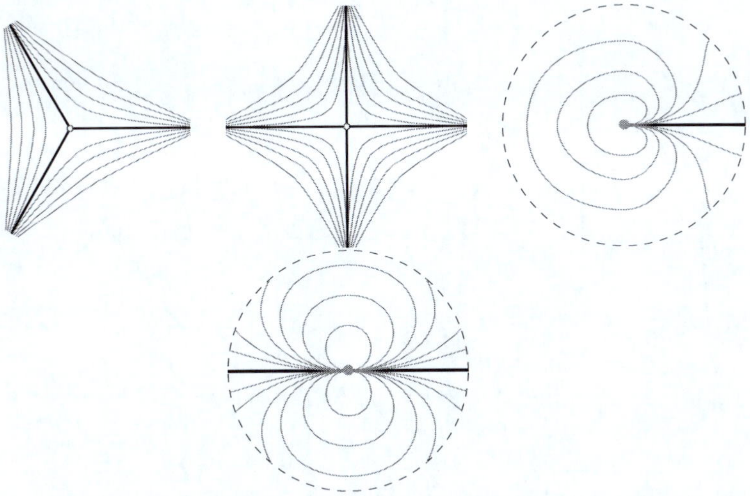

FIGURE 6.1. Foliations near classical type of zeroes/poles

points on ∂_P. We denote by S^ϕ the *(open) marked surface* associated to (S, ϕ), that consists of the surface S^ϕ together with those marked points. The *decorated marked surface* $S^\phi_{\mathrm{Zero}(\phi)}$ associated to (S, ϕ) is obtained from S^ϕ by decorating the set $\mathrm{Zero}(\phi)$ of zeroes of ϕ.

EXAMPLE 6.5. Definition 6.4 fits perfectly with triangulated surfaces in the theory of cluster algebras [**FST**]. For instance, the foliation of a quadratic differential on $\mathbb{P}^1$ with type $(1, 1, 1, -7)$, i.e., it has three simple zeroes and one order 7 pole, is shown in Figure 6.2. The left picture is the real blow-up part and the right picture is the neighbourhood of the pole. Note that, in the pictures,

- the blue vertices are poles or marked points;
- the red vertices are simple zeroes;
- the green arcs are geodesics;
- the black arcs are separating trajectories;
- the red arcs are the saddle connections in the horizontal strips.

Observe that the isotopy classes of separating trajectories in the left picture of Figure 6.2 form a *WKB-triangulation* of the marked surface (a pentagon in this case). In general, such a WKB-triangulation will be formed provided the GMN differential is saddle-free.

Let $\mathbf{S}^3_\Delta$ be a decorated marked surface with $\mathbf{M}$ the set of open marked points and Δ the set of decorations in the interior. Its numerical data satisfies (cf. [**Q2**, Def. 3.1])

$$|\Delta| = 4g + 2|\partial \mathbf{S}| + |M| - 4 \quad (\geq 1), \tag{6.2}$$

where g is the genus of $\mathbf{S}$.

DEFINITION 6.6. A $\mathbf{S}^3_\Delta$*-framed quadratic differential* (S, ϕ, ψ) consists of a GMN differential ϕ on a Riemann surface S together with a diffeomorphism $\psi \colon \mathbf{S}^3_\Delta \to S^\phi_{\mathrm{Zero}(\phi)}$ preserving the marked/decorating points setwise. Two $\mathbf{S}^3_\Delta$-framed quadratic differentials (S_1, ϕ_1, ψ_1) and (S_2, ϕ_2, ψ_2) are equivalent if

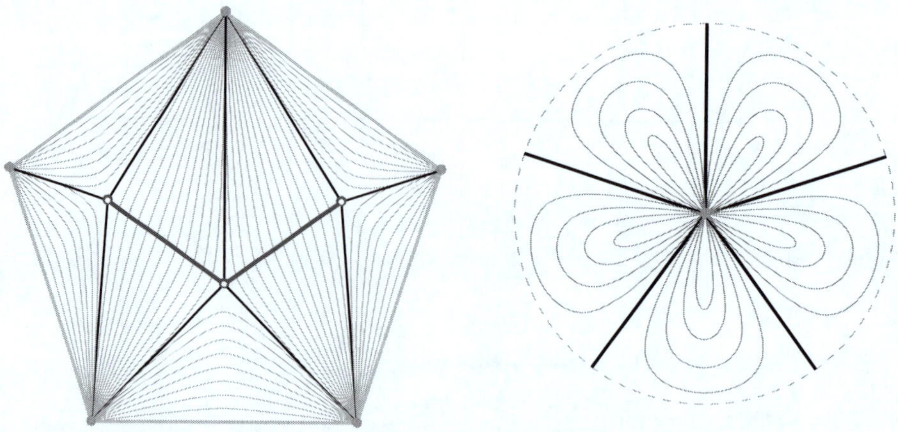

FIGURE 6.2. The foliation of a GMN differential on $\mathbb{P}^1$ for the A_2 case

- there exists a biholomorphism $f\colon \mathrm{S}_1 \to \mathrm{S}_2$ such that $f^*(\phi_2) = \phi_1$ and,
- $\psi_2^{-1} \circ f_* \circ \psi_1 \in \mathrm{Homeo}_0(\mathbf{S}^3_\Delta)$, where $f_*\colon (\mathrm{S}_1)^{\phi_1}_{\mathrm{Zero}(\phi_1)} \to (\mathrm{S}_2)^{\phi_2}_{\mathrm{Zero}(\phi_2)}$ is the induced homeomorphism. Recall that Homeo_0 consists of isotopy classes of homeomorphisms that are isotopy to identity.

We denote by $\mathrm{FQuad}_3(\mathbf{S}^3_\Delta)$ the moduli space of $\mathbf{S}^3_\Delta$-framed quadratic differentials. For $\mathbf{S}^3_\Delta$, there is a canonical associated Calabi-Yau-3 category $\mathcal{D}_3(\mathbf{S}^3_\Delta)$, which can be constructed as follows.

- Take any triangulation T of $\mathbf{S} = \mathrm{S}^\phi$ (or a WKB-triangulation mentioned above) and consider the corresponding quiver with potential (Q_T, W_T), where
  - the vertices of Q_T correspond to the open arcs in T;
  - the arrows of Q_T correspond to the angles of triangles in T;
  - the terms/cycles of W_T correspond to triangles in T.
- Let Γ_T be the Ginzburg dg algebra of (Q_T, W_T) and then $\mathcal{D}_3(\mathbf{S}^3_\Delta)$ is the finite dimensional derived category of Γ_T.

Recall that a triangulation is a maximal compatible collection of open arcs on $\mathbf{S}$. Moreover, [**BQZ**, Thm. A] shows that $\mathcal{D}_3(\mathbf{S}^3_\Delta)$ is unique (up to natural transformation) in a suitable sense. Furthermore, $\mathcal{D}_3(\mathbf{S}^3_\Delta)$ can be embedded into the derived Fukaya category of a symplectic 3-fold, constructed from fibrations of $\mathbf{S}$ via a quadratic differential [**S**].

Let $\widetilde{\mathbf{S}^3_\Delta}$ be the branched double cover (*spectral cover*) of $\mathbf{S}^3_\Delta$, branching at Δ. We finish this subsection with the following result, whose original version/idea is due to [**BS**], while the quoted upgraded version is taken from [**KQ2**].

THEOREM 6.7. [**BS, KQ2**] *There is an isomorphism*

$$\chi_3\colon \widehat{\mathbf{H}}(\mathbf{S}^3_\Delta) \to \mathrm{K}(\mathcal{D}_3(\mathbf{S}^3_\Delta)), \tag{6.3}$$

where the hat homology $\widehat{\mathbf{H}}(\mathbf{S}^3_\Delta) = \mathbf{H}_1(\widetilde{\mathbf{S}^3_\Delta}; \mathbb{Z})^-$ *is the anti-invariant part of* $\mathbf{H}_1(\widetilde{\mathbf{S}^3_\Delta}; \mathbb{Z})$ *with respect to the covering involution (cf. Definition* 7.11 *for the precise definition in the general setting). Moreover, there is an isomorphism* Ξ_3 *of complex manifolds that fits into the commutative diagram*

$$\begin{array}{ccc} \mathrm{FQuad}_3^\circ(\mathbf{S}^3_\Delta) & \xrightarrow{\Xi_3} & \mathrm{Stab}^\circ \mathcal{D}_3(\mathbf{S}^3_\Delta) \\ \Pi_3 \downarrow & & \downarrow \mathcal{Z}_3 \\ \mathrm{Hom}(\widehat{\mathbf{H}}(\mathbf{S}^3_\Delta), \mathbb{C}) & \xrightarrow{\chi_3} & \mathrm{Hom}(\mathrm{K}(\mathcal{D}_3(\mathbf{S}^3_\Delta)), \mathbb{C}), \end{array} \tag{6.4}$$

such that the period map $\Pi_3\colon \phi \mapsto \int \sqrt{\phi}$ *becomes the map* $\mathcal{Z}_3$ *(i.e., central charge). More precisely,* (1.3) *holds for*

$$X = X_3\colon\ \mathrm{CA}(\mathbf{S}^3_\Delta) \to \mathrm{Sph}\,\mathcal{D}_3(\mathbf{S}^3_\Delta)/[1] \tag{6.5}$$

in [**Q2**, Thm. 6.6], *where* $\mathrm{CA}(\mathbf{S}^3_\Delta)$ *the set of closed arcs in* $\mathbf{S}^3_\Delta$ *and* $\mathrm{Sph}\,\mathcal{D}_3(\mathbf{S}^3_\Delta)$ *the set of reachable spherical objects in* $\mathcal{D}_3(\mathbf{S}^3_\Delta)$.

Furthermore, for $\sigma = \Xi_3(\phi)$, *the map* X_3 *induces one-one correspondence between saddle connections of* ϕ *on* $\mathbf{S}^3_\Delta$ *and* σ*-semistable objects in* $\mathcal{D}_3(\mathbf{S}^3_\Delta)$.

Here, $^\circ$ in FQuad_3° and Stab° means taking the *principal component* (cf. [**KQ2**, § 4]).

We also have the moduli space of $\mathbf{S}^3$-framed quadratic differentials such that

$$\mathrm{FQuad}(\mathbf{S}^3) = \mathrm{FQuad}^\circ(\mathbf{S}^3_\Delta)/\,\mathrm{BT}(\mathbf{S}^3_\Delta) \cong \mathrm{Stab}\,\mathcal{D}_3(\mathbf{S}^3_\Delta)/\,\mathrm{ST}\,\mathcal{D}_3(\mathbf{S}^3_\Delta), \tag{6.6}$$

noticing that (6.5) induces the isomorphism between $\mathrm{BT}(\mathbf{S}^3_\Delta)$ and $\mathrm{ST}\,\mathcal{D}_3(\mathbf{S}^3_\Delta)$, cf. [**Q2**, Thm.1], which is the Calabi-Yau-3 version of (5.8).

6.3. Exponential type singularities in the Calabi-Yau-∞ setting

DEFINITION 6.8. [**HKK**] The *exponential type singularity* p of index $(k, l) \in \mathbb{Z}_{>0} \times \mathbb{Z}$ of a quadratic differential ϕ on a Riemann surface S is a singularity p with local coordinate as

$$e^{z^{-k}} z^{-l} g(z)\, \mathrm{d}\, z^{\otimes 2} \tag{6.7}$$

for some non-zero holomorphic function $g(z)$.

A *flat surface* (S, ϕ) is a Riemann surface S together with quadratic differential ϕ whose singularities are all of exponential type as above.

By [**HKK**, Prop. 2.5], for a singularity p in S^ϕ with local coordinate (6.7), there are exactly k additional points, with respect to the ϕ-metric completion in the neighborhood of p; all of which are ∞-angle singularities, also known as the conical singularity with infinity angle described as below.

DEFINITION 6.9. Consider the spaces

$$C_N\colon = \mathbb{Z}_N \times \mathbb{R} \times \mathbb{R}_{\geq 0}/\sim, \qquad (k, x, 0) \sim (k+1, -x, 0) \text{ for } x \leq 0 \tag{6.8}$$

for $N \in \mathbb{Z}_{>0}$ and

$$C_\infty\colon = \mathbb{Z} \times \mathbb{R} \times \mathbb{R}_{\geq 0}/\sim, \qquad (k, x, 0) \sim (k+1, -x, 0) \text{ for } x \leq 0. \tag{6.9}$$

The origin $0 = (0, 0, 0) \sim (k, 0, 0)$ is a distinguished point in these spaces. The standard flat structure of $\mathbb{R}^2$ makes $C_? \setminus \{0\}$ a smooth flat surface with metric completion $C_?$. The origin is a *conical singularity with* $N\pi$ *angle* for C_N, $N \neq 2$ and is a *conical singularity with infinity angle* for C_∞.

Note that the conical singularities with $N\pi$ angle are exactly classical type zeroes of order $N-2$ mentioned above, for $N \geq 3$. For GMN differentials (i.e., in BS' setting), we always have $N = 3$. We will discuss the $N \geq 3$ case in Section 9. Furthermore, the local behavior in the origin of C_∞ is similar to zeroes instead of poles since a trajectory attending to such points has finite length.

Let S_{sg} be the set of singularities of the flat surface S^ϕ (where the smooth part of S^ϕ is just S). We have the following notions.

- A *saddle connection* is a maximal geodesic converging towards closed points in S_{sg} (i.e., zeroes or points in the metric completion) in both directions. Note that saddle trajectories are saddle connections which are horizonal. Also, the endpoints are not required to be distinct.
- The *core* $\mathrm{Core}(\phi)$ is the convex hull of S_{sg} by [**HKK**, Prop. 2.2], which contains all saddle connections. Moreover, $\mathrm{Core}(\phi)$ is a deformation retract of S^ϕ provided that each point on S^ϕ has finite distance to $\mathrm{Core}(\phi)$ (with respect to the ϕ-metric), cf. [**HKK**, Prop. 2.3].

DEFINITION 6.10. The *graded marked surface* S^ϕ associated to a flat surface (S, ϕ) is the *real blow-up* of S at all the singularities of ϕ. So

- Each boundary component of S^ϕ corresponds to the blow-up of a singularity p of exponential type (k, l).
- The closed marked points correspond to the additional k points with respect to ϕ-metric completion mentioned above.
- The open marked points correspond to the directions between closed marked points (hence are dual to closed marked points on the boundaries).
- The grading is given by the horizontal foliation.

REMARK 6.11. In fact, any graded marked surface $(\mathbf{S}, \mathbf{M}, \mathbf{Y}, \lambda)$ in Section 3 is the real blow-up of some flat surfaces with conical singularities of infinity angle.

DEFINITION 6.12. Let $\mathbf{S}^\lambda$ be a graded marked surface with the numerical data

$$\mathfrak{N}(\mathbf{S}^\lambda) = (g, b; \mathbf{k}, \mathbf{l}; \mathrm{LP}_g).$$

A $\mathbf{S}^\lambda$*-framed quadratic differential* (S, ϕ, h) on $\mathbf{S}^\lambda$ consists of a quadratic differential ϕ on some Riemann surface S together with a diffeomorphism $h \colon \mathbf{S}^\lambda \to \mathrm{S}^\phi$ such that the numerical data of the grading $h^*(\phi)$ on $\mathbf{S}^\lambda$ (the pull-back of the ϕ-induced grading on S) equals $\mathfrak{N}(\mathbf{S}^\lambda)$. Two $\mathbf{S}^\lambda$-framed quadratic differentials $(\mathrm{S}_i, \phi_i, h_i)$ are equivalent if

- there exists a biholomorphism $f \colon \mathrm{S}_1 \to \mathrm{S}_2$ such that $f^*(\phi_2) = \phi_1$;
- $h_2^{-1} \circ f_* \circ h_1 \in \mathrm{Diff}_0(\mathbf{S}^\lambda)$, where $f_* \colon \mathrm{S}_1^{\phi_1} \to \mathrm{S}_2^{\phi_2}$ is the induced diffeomorphism.

Denote by $\mathrm{FQuad}_\infty(\mathbf{S}^\lambda)$ the space of $\mathbf{S}^\lambda$-framed quadratic differentials on $\mathbf{S}^\lambda$. For simplicity, we will only mention a quadratic differential ϕ on $\mathbf{S}^\lambda$ (omitting S and h).

Now we recall the main result in [**HKK**], where the surjection of the Ξ_∞ is shown by [**Ta**].

THEOREM 6.13. [**HKK, Ta**] *There is the following identification*

$$\chi_\infty \colon \mathbf{H}_1(\mathbf{S}^\lambda, \partial\mathbf{S}^\lambda; \mathbb{Z}_{\mathrm{sp}}) \to \mathrm{K}(\mathcal{D}_\infty(\mathbf{A})), \tag{6.10}$$

where $\mathbb{Z}_{\mathrm{sp}} = \mathbb{Z} \otimes_{\mathbb{Z}_2} \mathrm{sp}$ *for* sp *being the canonical double cover of* $\mathbf{S}^\lambda$. *Moreover, there is an isomorphism* Ξ_∞ *of complex manifolds that fits into the commutative diagram*

$$\begin{array}{ccc} \mathrm{FQuad}_\infty(\mathbf{S}^\lambda) & \xrightarrow{\Xi_\infty} & \mathrm{Stab}\,\mathcal{D}_\infty(\mathbf{A}) \\ \downarrow{\scriptstyle \Pi_\infty} & & \downarrow{\scriptstyle \mathcal{Z}_\infty} \\ \mathrm{Hom}(\mathbf{H}_1(\mathbf{S}^\lambda, \partial\mathbf{S}^\lambda; \mathbb{Z}_{\mathrm{sp}}), \mathbb{C}) & \xrightarrow{\chi_\infty} & \mathrm{Hom}(\mathrm{K}(\mathcal{D}_\infty(\mathbf{A})), \mathbb{C}), \end{array} \tag{6.11}$$

such that the period map $\Pi_\infty \colon \phi \mapsto \int \sqrt{\phi}$ *becomes the map* $\mathcal{Z}_\infty$ *(i.e., central charge). More precisely,* (1.3) *holds for* $X = \widetilde{X}_\infty$ *in* (3.8) *in Theorem* 3.7.

Furthermore, for $\widehat{\sigma} = \Xi_\infty(\phi)$, *the map* $\widetilde{X}_\infty$ *induces a one-one correspondence between saddle connections of* ϕ *on* $\mathbf{S}^\lambda$ *and* $\widehat{\sigma}$*-semistable objects in* $\mathcal{D}_\infty(\mathbf{A})$.

CHAPTER 7

q-Deformation of quadratic differentials

In this section, we fix a complex number $s \in \mathbb{C}$ with $\operatorname{Re}(s) > 2$ and set $q_s = e^{\mathbf{i}\pi s}$.

7.1. Multi-valued quadratic differentials

Now we introduce the q-deformed generalization of GMN-differentials in Section 6.2. We first define q_s-quadratic differentials on a disk. Fix the following notations:

- $D := \{z \in \mathbb{C} \mid |z| < 1\}$ is the unit disk;
- $D^* := D \setminus \{0\}$ is the punctured unit disk;
- $\widetilde{D^*}$ is the universal cover of D^*.

In fact, $\widetilde{D^*}$ is isomorphic to the half plane $\{w \in \mathbb{C} \mid \operatorname{Re}(w) < 0\}$ with covering map

$$\begin{array}{ccc} \widetilde{D^*} & \longrightarrow & D^* \\ w & \mapsto & e^w. \end{array}$$

Consider a holomorphic quadratic differential

$$e^{(ks+l+2)w} \, \mathrm{d}\, w^{\otimes 2}$$

on $\widetilde{D^*}$, where $k, l \in \mathbb{Z}$. Set $z = e^w$. Then $\mathrm{d}\, z = e^w \,\mathrm{d}\, w$ and one can view

$$\xi = z^{ks+l} \, \mathrm{d}\, z^{\otimes 2}$$

as a multi-valued quadratic differential on D ramified at 0. Let γ be a counterclockwise loop around 0. Since the action of the deck transformation group $\pi_1(D^*) \cong \langle \gamma \rangle$ on $\widetilde{D^*}$ is given by $\gamma(w) = w + 2\pi\mathbf{i}$, the monodromy of ξ is given by $\gamma^*\xi = q_s^{2k}\xi$.

Now we turn to general case. Let S be a compact Riemann surface and $\mathfrak{R} \subset \mathrm{S}$ a set of finite points. Denote by

$$\Pi \colon \widetilde{\mathrm{S}^\phi} \to \mathrm{S}^\phi \tag{7.1}$$

the universal cover of $\mathrm{S}^\phi := \mathrm{S} \setminus \mathfrak{R}$. In our setup, a *multi-valued quadratic differential on* S *ramified at* $\mathfrak{R}$ refers to a holomorphic quadratic differential on $\widetilde{\mathrm{S}^\phi}$. The fundamental group $\pi_1(\mathrm{S}^\phi, *)$ acts on $\widetilde{\mathrm{S}^\phi}$ as deck transformations. We denote $\gamma^*\xi$ the pull-back of a multi-valued quadratic differential ξ via the action of a deck transformation $\gamma \in \pi_1(\mathrm{S}^\phi, *)$.

DEFINITION 7.1. A *q_s-quadratic differential ξ on* S *ramified at* $\mathfrak{R}$ is a multi-valued quadratic differential on S ramified at $\mathfrak{R}$ such that, for any $\gamma \in \pi_1(\mathrm{S}^\phi, *)$,

$$\gamma^*\xi = q_s^{2m(\gamma)}\xi \tag{7.2}$$

for some $m(\gamma) \in \mathbb{Z}$.

For a point $p \in \mathrm{S}$, let D_p be a small disk (neighbourhood) centered at p with coordinate z. Similarly as above, $D_p^* = D_p \setminus \{p\}$ and $\widetilde{D_p^*}$ its universal cover. Then $D_p^* \subset \mathrm{S}^\phi$ for any $p \in \mathfrak{R}$ and the pull-back

$$\Pi^{-1}(D_p^*) = \coprod_\alpha \widetilde{D_p^*}^\alpha \tag{7.3}$$

is the disjoint unions of countable copies of $\widetilde{D_p^*}$. We say a q_s-quadratic differential ξ has the local form

$$\xi = c\, z^{k_p s + l_p}\, \mathrm{d}\, z^{\otimes 2} \tag{7.4}$$

on D_p if

$$\xi = c\, e^{(k_p s + l_p + 2)w}\, \mathrm{d}\, w^{\otimes 2}$$

on some connected component $\widetilde{D_p^*}^0 \subset \Pi^{-1}(D_p^*)$, for some $c \in \mathbb{C}$ and $k_p, l_p \in \mathbb{Z}$. Note that in such a case, by Definition 7.1, ξ can be written as

$$\xi = c\, q_s^{2m_\alpha}\, e^{(k_p s + l_p + 2)w}\, \mathrm{d}\, w^{\otimes 2},$$

for some $m_\alpha \in \mathbb{Z}$, on any connected component $\widetilde{D_p^*}^\alpha \subset \Pi^{-1}(D_p^*)$.

Now we introduce the notion of zeroes and poles for q_s-quadratic differentials.

DEFINITION 7.2. Let ξ be a q_s-quadratic differential and assume that it has the local form (7.4) on D_p for some $p \in \mathfrak{R}$, where $c \in \mathbb{C}$ and $k_p, l_p \in \mathbb{Z}$. We call p

- a *zero of order* $(k_p s + l_p)$ if $\mathrm{Re}(k_p s + l_p) > 0$;
- a *pole of order* $-(k_p s + l_p)$ if $\mathrm{Re}(k_p s + l_p) < 0$.

And a pole p of ξ is a *higher order pole* if $\mathrm{Re}(k_p s + l_p) < -2$. Moreover,

- a zero p of ξ is called s-*simple* if ξ has the local form

$$\xi = c z^{s-2}\, \mathrm{d}\, z^{\otimes 2}; \tag{7.5}$$

- a pole p of ξ is called an s-*pole of type* $(k, l) \in \mathbb{Z}^2$ if ξ has the local form

$$\xi = c z^{-k(s-2)-l}\, \mathrm{d}\, z^{\otimes 2}. \tag{7.6}$$

Associating to a q_s-quadratic differential ξ, there is a multi-set of pairs of integers

$$(\mathbf{k}, \mathbf{l})\colon = \{(k_1, l_1), \ldots, (k_b, l_b)\} \tag{7.7}$$

defined via gathering all types of s-poles of ξ, where b is the number of poles of ξ. We call the multi-set $(\mathbf{k}, \mathbf{l})$ the s-*polar type of* ξ.

REMARK 7.3. If $s \in \mathbb{Q}$, the pair of integers (k_i, l_i) is not uniquely determined by the local form ξ around the corresponding s-pole. Therefore the s-polar type $(\mathbf{k}, \mathbf{l})$ for ξ carries the additional information.

Recall that a q_s-quadratic differential ξ defines the monodromy representation

$$\begin{array}{rcl} \rho_s\colon \pi_1(\mathrm{S}^\phi, *) & \to & \mathbb{C}^* \\ \gamma & \mapsto & q_s^{2m(\gamma)}, \end{array} \tag{7.8}$$

where $m(\gamma)$ is given by (7.2).

ASSUMPTION 7.4. *For a q_s-quadratic differential ξ, there is a simply connected open domain $U \subset \mathrm{S}$ such that:*

- *U contains all ramification points $\mathfrak{R}$, i.e., $\mathfrak{R} \subset U$;*

- *the induced monodromy representation* $\rho_s \circ \iota_* \colon \pi_1(\mathrm{S} \setminus U, *) \to \mathbb{C}^*$, *via the inclusion* $\iota \colon \mathrm{S} \setminus U \to \mathrm{S}^\phi$, *is trivial.*

In what follows, we first introduce a special class of q_s-quadratic differentials and then consider their moduli spaces.

DEFINITION 7.5. Let g be a non-negative integer and $(\mathbf{k}, \mathbf{l})$, as in (7.7), a multi-set of pairs of integers satisfying $k_i > 0$ and (3.6). A *CY-s type q_s-quadratic differential* ξ on genus g Riemann surfaces S is a q_s-quadratic differential satisfying:

- all zeroes are s-simple;
- the number of s-poles is b and the s-polar type is $(\mathbf{k}, \mathbf{l})$;
- all poles are higher order s-poles, namely,

$$\mathrm{Re}(k_i(s-2) + l_i) > 2, \qquad \forall i = 1, \ldots, b; \tag{7.9}$$

- Assumption 7.4.

We fix the following convention of notations.

- $\mathrm{Zero}(\xi) = \{Z_1, \ldots, Z_\aleph\}$ is the set of $\aleph$ s-simple zeroes.
- $\mathrm{Pol}(\xi) = \{p_1, \ldots, p_b\}$ is the set of s-poles with p_i of type (k_i, l_i).

In the setting above, the set of ramification points is $\mathfrak{R} = \mathrm{Zero}(\xi) \bigcup \mathrm{Pol}(\xi)$.

The next statement clarifies the relationship between the number $\aleph$ of s-simple zeroes of a CY-s type q_s-quadratic differential and its s-polar type.

LEMMA 7.6. *Let ξ be a CY-s type q_s-quadratic differential of s-polar type $(\mathbf{k}, \mathbf{l})$ as in (7.7). Then we have*

$$\aleph = \sum_{i=1}^{b} k_i. \tag{7.10}$$

PROOF. For the case $s \notin \mathbb{Q}$, denote by γ_j an anticlockwise loop around Z_j and by δ_i an anticlockwise loop around p_i. Then Assumption 7.4 implies that

$$\prod_{j=1}^{\aleph} \rho_s(\gamma_j) \cdot \prod_{i=1}^{b} \rho_s(\delta_i) = 1.$$

Since $\rho_s(\gamma_j) = q_s^2$ and $\rho(\delta_i) = q_s^{-2k_i}$, the above equality yields

$$\left(2\aleph - 2\sum_{i=1}^{b} k_i \right) s \in 2\mathbb{Z}.$$

Therefore, by the fact that $s \notin \mathbb{Q}$, we have (7.10).

For the case $s \in \mathbb{Q}$, we can assume $s = j/m$ for some $j, m \in \mathbb{Z}$. Note that ξ^m is a single-valued meromorphic section of $\omega_S^{\otimes 2m}$ with $\aleph$ zeroes of order $m(s-2)$ and b poles of orders $m(k_i(s-2) + l_i)$, for $1 \le i \le b$. By the Riemann-Roch theorem, we obtain

$$m(4g-4) = m\aleph(s-2) - m\sum_{i=1}^{b}(k_i(s-2) + l_i)$$

which implies (7.10) noticing (3.6). This completes the proof. □

REMARK 7.7. When $s = 3$, we can recover the GMN differential on Riemann surfaces in Definition 6.4 ([**BS**]). For instance, when S is $\mathbb{P}^1$ with numerical data (i.e., the normal A_2 case) (cf. Example 3.3)

$$g = 0,\ b = 1,\ (k, l) = (3, 4),$$

we will get the corresponding CY-3 A_2 example in Example 6.5 with singularity type $(1, 1, 1, -7)$, whose foliation as shown in Figure 6.2.

7.2. Log surfaces

Continuing Section 7.1, we consider a CY-s type q_s-quadratic differential ξ on a genus g Riemann surface S with an s-polar type $(\mathbf{k}, \mathbf{l})$. As before, denote by γ_j the loop around Z_j and by δ_i the loop around p_i in the simply connected domain $U \subset \mathrm{S}$.

Define the representation of $\pi_1(\mathrm{S}^\phi, *)$ on the infinite cyclic group $\langle q \rangle \cong \mathbb{Z}$ as follows. To start with, we have a representation

$$\underline{\rho}\colon \pi_1(U \setminus \mathfrak{R}, *) \to \langle q \rangle$$

defined by setting $\underline{\rho}(\gamma_j) := q$ and $\underline{\rho}(\delta_i) := q^{-k_i}$. Note that for a loop $\Gamma \subset U$ encircling all points in $\mathfrak{R}$, we have $\underline{\rho}(\Gamma) = 1$. Then we can extend $\underline{\rho}$ to the representation

$$\rho\colon \pi_1(\mathrm{S}^\phi, *) \to \langle q \rangle, \tag{7.11}$$

such that

- the restriction of ρ on $U \setminus \mathfrak{R}$ is $\underline{\rho}$,
- $\rho(\gamma) = 1$ for any loop $\gamma \in \pi_1(\mathrm{S} \setminus U)$.

Using the representation ρ, we introduce logarithmic surfaces.

DEFINITION 7.8. The *logarithmic (log) surface* $\log \mathrm{S}^\phi$ *associated to* ξ is defined to be the covering space $\pi\colon \log \mathrm{S}^\phi \to \mathrm{S}^\phi$ corresponding to the representation ρ. Since ξ is a single-valued quadratic differential on $\log \mathrm{S}^\phi$, the latter has a natural grading in the sense of Definition 3.1 given by the horizontal foliation of ξ. The infinite cyclic group $\langle q \rangle$ acts on $\log \mathrm{S}^\phi$ as deck transformations.

We end this section with one more remark.

REMARK 7.9. For an s-simple zero Z_j with loop γ_j around it, we have $\rho(\gamma_j) = q$. Taking $p = Z_j$, we know that $\Pi^{-1}(D_p^*)$ in (7.3) consists of just one copy of $\widetilde{D_p^*}$. Similarly, for an s-pole p_i of type (k_i, l_i) with loop δ_i around it, we have $\rho(\delta_i) = k_i$. Taking $p = p_i$, we know that $\Pi^{-1}(D_p^*)$ in (7.3) becomes

$$\Pi^{-1}(D_p^*) = \coprod_{i=1}^{k_i} \widetilde{D_p^*}^{\,i} \tag{7.12}$$

consisting of k_i copy of $\widetilde{D_p^*}$. The deck transformation q maps $\widetilde{D_p^*}^{\,i}$ to $\widetilde{D_p^*}^{\,i+1}$, where the subscript $i, i+1$ is considered modulo k_i.

7.3. Hat homology groups

Let $\log \mathrm{S}^\phi$ be the log surface associated to a q_s-quadratic differential ξ on S and

$$\psi_s := \sqrt{\xi} \tag{7.13}$$

the square root of ξ. Note that ψ_s is a double-valued holomorphic 1-form on $\log \mathrm{S}^\phi$ and it defines the monodromy representation

$$\begin{array}{rcl} \rho^{\frac{1}{2}}\colon \pi_1(\mathrm{S}^\phi, *) & \to & \langle q, -q\rangle \\ \gamma & \mapsto & \pm q^{m(\gamma)}, \end{array} \tag{7.14}$$

where $\gamma^*\psi_s = \pm q_s^{m(\gamma)}\psi_s$. The image of $\rho^{\frac{1}{2}}$ is given by

$$\operatorname{Im}\rho^{\frac{1}{2}} = \langle q\rangle \quad \text{or} \quad \langle q, -q\rangle.$$

Now we consider the Riemann surface associated to ψ_s.

DEFINITION 7.10. The *spectral cover*

$$\widehat{\mathrm{sp}}\colon \log \widehat{\mathrm{S}}^\circ \to \log \mathrm{S}^\phi$$

is defined to be the covering space corresponding to the representation $\rho^{\frac{1}{2}}$.

The holomorphic 1-form ψ_s is single-valued on $\log \widehat{\mathrm{S}}^\circ$. Note that

- $\log \widehat{\mathrm{S}}^\circ$ is a double cover of $\log \mathrm{S}^\phi$ if $\operatorname{Im}\rho^{\frac{1}{2}} = \langle q, -q\rangle$;
- $\log \widehat{\mathrm{S}}^\circ = \log \mathrm{S}^\phi$ if $\operatorname{Im}\rho^{\frac{1}{2}} = \langle q\rangle$.

Let $\widehat{\mathrm{S}}^\circ$ be the covering space $\mathrm{sp}\colon \widehat{\mathrm{S}}^\circ \to \mathrm{S}^\phi$ corresponding to the representation

$$\pi_1(\mathrm{S}^\phi, *) \to \{\pm 1\},$$

which is obtained by setting $q \mapsto 1$ in the definition of $\rho^{\frac{1}{2}}$. The above constructions give rise to the following commutative diagram:

$$\begin{array}{ccc} \log \widehat{\mathrm{S}}^\circ & \xrightarrow{\ \widehat{\pi}\ } & \widehat{\mathrm{S}}^\circ \\ {\scriptstyle \widehat{\mathrm{sp}}}\downarrow & & \downarrow{\scriptstyle \mathrm{sp}} \\ \log \mathrm{S}^\circ & \xrightarrow{\ \pi\ } & \mathrm{S}^\phi. \end{array}$$

Moreover, the double cover $\mathrm{sp}\colon \widehat{\mathrm{S}}^\circ \to \mathrm{S}^\phi$ can be extended to a branched double cover

$$\mathrm{Sp}\colon \widehat{\mathrm{S}} \to \mathrm{S},$$

branching at the poles of ξ with odd winding numbers (equivalently, with odd l_i).

To introduce an appropriate integration of ψ_s on $\log \widehat{\mathrm{S}}^\circ$, we define the following homology group.

DEFINITION 7.11. The *hat homology group* of ξ is defined as follows.

- If $\log \widehat{\mathrm{S}}^\circ = \log \mathrm{S}^\phi$, then
$$\widehat{\mathbf{H}}(\xi)\colon = \mathbf{H}_1(\log \mathrm{S}^\phi; \mathbb{Z}).$$
- If $\log \widehat{\mathrm{S}}^\circ$ is a double cover of $\log \mathrm{S}^\phi$, denote by τ the covering involution and
$$\widehat{\mathbf{H}}(\xi) := \mathbf{H}_1(\log \mathrm{S}^\phi; \mathbb{Z})^-,$$

where $\mathbf{H}_1(\log \mathrm{S}^{\phi};\mathbb{Z})^-$ is the τ-anti-invariant part, i.e., consists of $\gamma \in \mathbf{H}_1(\log \widehat{\mathrm{S}}^\circ;\mathbb{Z})$ satisfying $\tau_*\gamma = -\gamma$. The group $\widehat{\mathbf{H}}(\xi)$ has an $R = \mathbb{Z}[q^{\pm 1}]$-module structure induced from the action of the deck transformation group $\langle q, -q\rangle$ (or $\langle q\rangle$) on $\log \widehat{\mathrm{S}}^\circ$.

DEFINITION 7.12. The *period* of ξ is the integration of ψ_s on cycles in $\widehat{\mathbf{H}}(\xi)$, which is an R-linear map:

$$\begin{aligned} Z_\xi \colon \widehat{\mathbf{H}}(\xi) &\to \mathbb{C}_s \\ \gamma &\mapsto \int_\gamma \psi_s. \end{aligned} \tag{7.15}$$

Note that the construction of ρ implies that

$$q^*\psi_s = q_s\psi_s \tag{7.16}$$

and hence

$$\int_{q_*\gamma} \psi_s = \int_\gamma q^*\psi_s = q_s \int_\gamma \psi_s$$

for any cycle γ in $\widehat{\mathbf{H}}(\xi)$.

The rest of the subsection is devoted to the calculation of the rank of $\widehat{\mathbf{H}}(\xi)$.

PROPOSITION 7.13. *Let ξ be a CY-s type q_s-quadratic differential of genus g and s-polar type $(\mathbf{k},\mathbf{l})$ as in (7.7). Then the associated hat homology group $\widehat{\mathbf{H}}(\xi)$ is a free R-module of rank*

$$n = 2g - 2 + b + \sum_{i=1}^{b} k_i.$$

To prove the statement, we prepare several lemmas first. Let $B^{(m)}$ be a bouquet of m circles joined at the point $*$, namely,

$$B^{(m)} = S^1 \vee \cdots \vee S^1.$$

The fundamental group of $B^{(m)}$ is a free group with m generators, i.e.,

$$\pi_1(B^{(m)}, *) \cong \langle \gamma_1, \ldots, \gamma_m\rangle,$$

where, for $1 \le i \le m$, each γ_i is a loop around the i-th circle of $B^{(m)}$.

Let

$$\rho \colon \pi_1(B^{(m)}, *) \to \langle q\rangle$$

be any given surjective group homomorphism and take the covering space

$$\pi \colon \widetilde{B}^{(m)} \to B^{(m)}$$

corresponding to ρ. Set $\widetilde{*} = \pi^{-1}(*)$.

LEMMA 7.14. *The relative homology group $\mathbf{H}_1(\widetilde{B}^{(m)}, \widetilde{*};\mathbb{Z})$ is a free R-module of rank m generated by the classes $[\widetilde{\gamma}_1], \ldots, [\widetilde{\gamma}_m]$, where $\widetilde{\gamma}_i$ is a lift of the loop γ_i around the i-th circle in $B^{(m)}$.*

PROOF. For each $1 \le i \le m$, take a lift $\widetilde{\gamma}_i$ of γ_i on $\widetilde{B}^{(m)}$. Then all other lifts are of the forms $(q_*)^{n_i}\widetilde{\gamma}_i$ for some $n_i \in \mathbb{Z}$. Due to the fact that any path which connects two points in $\widetilde{*}$ is homotopy equivalent to the composition of lifts of $\gamma_i^{\pm 1}$, any relative homology class in $\mathbf{H}_1(\widetilde{B}^{(m)}, \widetilde{*};\mathbb{Z})$ is a linear combination of classes $[\widetilde{\gamma}_i]$'s over R. It is clear that the set

$$\{[(q_*)^{n_i}\widetilde{\gamma}_i] \mid i = 1, \ldots, m, n_i \in \mathbb{Z}\}$$

is linearly independent. The claim follows. $\square$

LEMMA 7.15. *The homology group $\mathbf{H}_1(\widetilde{B}^{(m)};\mathbb{Z})$ is a free R-module of rank $m-1$.*

PROOF. Consider the exact sequence

$$0 \to \mathbf{H}_1(\widetilde{B}^{(m)};\mathbb{Z}) \to \mathbf{H}_1(\widetilde{B}^{(m)},\widetilde{*};\mathbb{Z}) \to \widetilde{\mathbf{H}}_0(\widetilde{*};\mathbb{Z}) \to 0,$$

where $\widetilde{\mathbf{H}}_0(\widetilde{*};\mathbb{Z})$ is the 0-th reduced homology group of $\widetilde{*}$. By definition and Lemma 7.14, we have $\widetilde{\mathbf{H}}_0(\widetilde{*};\mathbb{Z}) \cong R$ and $\mathbf{H}_1(\widetilde{B}^{(m)},\widetilde{*};\mathbb{Z}) \cong R^{\oplus m}$. Therefore the proof completes. □

PROPOSITION 7.16. *Let $\log \mathrm{S}^{\phi}$ be the logarithmic surface associated to a CY-s type q_s-quadratic differential of genus g and s-polar type $(\mathbf{k},\mathbf{l})$ as in* (7.7). *Then the homology group $\mathbf{H}_1(\log \mathrm{S}^{\phi};\mathbb{Z})$ is a free R-module of rank*

$$n = 2g - 2 + b + \sum_{i=1}^{b} k_i. \tag{7.17}$$

PROOF. By Lemma 7.6, we have

$$|\mathfrak{R}| = b + \aleph = b + \sum_{i=1}^{b} k_i.$$

Note that an oriented surface of genus g with $|\mathfrak{R}|$ points excluded is homotopy equivalent to a bouquet of $2g - 1 + |\mathfrak{R}|$ circles. Therefore $\mathrm{S}^{\phi} = \mathrm{S}\setminus\mathfrak{R}$ is homotopy equivalent to the bouquet $B^{(n+1)}$ and $\log \mathrm{S}^{\phi}$ is its covering that corresponds to ρ in (7.11). As $\rho(\gamma_j)q$ for loop γ_j around any s-zero Z_j, ρ is surjective. Hence the claim follows from Lemma 7.15. □

Proof of Proposition 7.13. If $\log\widehat{\mathrm{S}}^{\circ} = \log \mathrm{S}^{\phi}$, the statement follows from Proposition 7.16. Now consider the case when $\log\widehat{\mathrm{S}}^{\circ}$ is a double cover of $\log \mathrm{S}^{\phi}$. We first compute the rank of $\mathbf{H}_1(\log\widehat{\mathrm{S}}^{\circ};\mathbb{Z})$ over R. Recall that we have a underlying spectral/branched double cover $\widehat{\mathrm{sp}} : \widehat{\mathrm{S}}^{\circ} \to \mathrm{S}^{\phi}$ for $\widehat{\mathrm{S}}^{\circ} = \widehat{\mathrm{S}}\setminus\widehat{\mathfrak{R}}$ and $\widehat{\mathfrak{R}} = \widehat{\mathrm{sp}}^{-1}(\mathfrak{R})$. The Riemann-Hurwitz formula implies that

$$2\widehat{g} - 2 = 2(2g-2) + b_1,$$

where $\widehat{g}$ is the genus of $\widehat{\mathrm{S}}$ and b_1 is the number of odd integers in $\{l_1,\ldots,l_b\}$. In addition, we have

$$|\widehat{\mathfrak{R}}| = 2|\mathfrak{R}| - b_1.$$

Therefore the surface $\widehat{\mathrm{S}}^{\circ}$ is homotopy equivalent to the bouquet of m circles where

$$m = 2\widehat{g} - 1 + |\widehat{\mathfrak{R}}| = 4g - 3 + 2|\mathfrak{R}| = 2n + 1.$$

Then, by Proposition 7.16, the homology group $\mathbf{H}_1(\log\widehat{\mathrm{S}}^{\circ};\mathbb{Z})$ is a free R-module of rank $2n$. Since the τ-invariant part of $\mathbf{H}_1(\log\widehat{\mathrm{S}}^{\circ};\mathbb{Z})$ can be identified with $\mathbf{H}_1(\log \mathrm{S}^{\phi};\mathbb{Z})$, the rank of τ-anti-invariant part is given by

$$\mathrm{rank}_R\,\mathbf{H}_1(\log\widehat{\mathrm{S}}^{\circ};\mathbb{Z}) - \mathrm{rank}_R\,\mathbf{H}_1(\log \mathrm{S}^{\phi};\mathbb{Z}) = 2n - n = n.$$

□

7.4. Comparing log surfaces

Keep the notations as above, i.e. ξ is a CY-s type q_s-quadratic differential of genus g and s-polar type $(\mathbf{k}, \mathbf{l})$ on S ramified at

$$\mathfrak{R} = \mathrm{Zero}(\xi) \bigcup \mathrm{Pol}(\xi).$$

As in Section 7.2, near each s-simple zero $Z \in \mathrm{Zero}(\xi)$, we can take a disk $D_Z \subset \mathrm{S}$ with center Z and the restriction of covering map $\mathrm{sp}\colon \log \mathrm{S}^\phi \to \mathrm{S}^\phi$ on $D_Z^* = D_Z \setminus \{Z\}$ is the universal covering $\widetilde{D_Z^*} \subset \log \mathrm{S}^\phi$. By adding a point $\widetilde{Z}$ on $\widetilde{D_Z^*}$ as the fiber of Z, we can extend the covering $\widetilde{D_Z^*} \to D_Z^*$ to a branched $\mathbb{Z}$-covering $\widetilde{D_Z} \to D_Z$, branching at $\widetilde{Z}$.

LEMMA 7.17. *The construction above, i.e., adding $\widetilde{Z}$ can be viewed as the* metric completion *with respect to the ξ-metric on* $\log \mathrm{S}^\phi$.

PROOF. Take coordinate $w \in \widetilde{D_Z^*}$ and $z \in D_Z^*$ satisfying $z = e^w$. Then ξ on $\widetilde{D_Z^*}$ can be written as

$$\xi|_{\widetilde{D_Z^*}} = (e^w)^{s-2}(\mathrm{d}\, e^w)^{\otimes 2} = e^{sw}\, \mathrm{d}\, w^{\otimes 2}.$$

We identify

$$\widetilde{D_Z^*} \cong \{w \in \mathbb{C} \mid \mathrm{Re}\, w < 0\} \subset \mathbb{C} \cup \{\infty\} = \mathbb{CP}^1.$$

Then due to [**HKK**, Prop. 2.1 and Lemma 2.5], the point ∞ is an exponential type singularity and we can obtain a single additional point $\widetilde{Z}$ by the (partial) completion of $\widetilde{D_Z^*}$ with respect to the ξ-metric. □

Denote by $\log \mathrm{S}_\Delta$ the topological surface obtained from $\log \mathrm{S}^\phi$ by adding $\widetilde{Z_j}$ for each $Z_j \in \mathrm{Zero}(\xi)$. When there is no confusion, we will identify $\widetilde{Z_j}$ with Z and use the notation $\Delta = \mathrm{Zero}(\xi)$.

Next we consider an s-pole $p \in \mathrm{Pol}(\xi)$ of type (k, l) of ξ, and take a disk $D_p \subset \mathrm{S}$ with center p. Then the real blow-up of D_p at p is an annulus A_p obtained by replacing p with a boundary $\partial_p \simeq S^1$.

Now we consider the fiber of D_p^* in $\log \mathrm{S}^\phi$, as calculated in (7.12), consisting of k connected component $\widetilde{D_p^*}^i$. Corresponding to the real blow-up of D_p at p, we add a real line $\mathbb{R}$ on each $\widetilde{D_i^*}^i$ as the fiber of $\partial_p \subset A_p$ and extend the universal covering $\widetilde{D_p^*}^i \to D_p^*$ to the following one.

LEMMA 7.18. *The construction above leads to an extended universal covering*

$$\widetilde{D_p^*}^i \cup \mathbb{R} =: \widetilde{A_p}^i \longrightarrow A_p.$$

In the coordinate

$$\widetilde{D_p^*}^i \cong \{x + \mathbf{i}y \mid x < 0, y \in \mathbb{R}\},$$

the real line $\mathbb{R}$ corresponds to $\{-\infty + \mathbf{i}y \mid y \in \mathbb{R}\}$.

We call this construction an *real blow-up of* $\log \mathrm{S}^\phi$ at a pole p. Denote by $(\log \mathrm{S}^\phi)^\xi$ the surface obtained from $\log \mathrm{S}^\circ$ by performing real blow-up at all poles of ξ.

DEFINITION 7.19. The log surface $\log \mathrm{S}_\Delta^\xi$ associated to a CY-s type q_s-quadratic differential ξ on S is the topological surface obtained from $\log \mathrm{S}^\phi$ by adding back all zeroes in $\mathrm{Zero}(\xi)$ as above and performing real blow-up at all poles in $\mathrm{Pol}(\xi)$.

REMARK 7.20. We define the numerical data

$$\mathfrak{N}(\log \mathrm{S}_\Delta^\xi) = \mathfrak{N}(\mathrm{S}, \xi) = (g, b; \mathbf{k}, \mathbf{l}; \mathrm{LP}_g)$$

of $\log \mathrm{S}_\Delta^\xi$ as follows:

- The genus g and the number b of boundary components are inherited from S.
- $(\mathbf{k}, \mathbf{l})$ is given by the polar type of ξ.
- The Lekili-Polishchuk data LP_g is defined the same way as in Definition 3.2, via the winding numbers on S with respect to ξ. Note that, when choosing (e.g. non-separating) curves, one needs to avoid the simply connected domain U in Assumption 7.4.

Then we have the following, which basically says that the winding number of an exponential pole of polar type (k, l) with local coordinate (6.7) matches the winding number of a CY-s pole of polar type (k, l) with local coordinate (7.6). Cf. Figure 8.1 for illustration for the case when $(k, l) = (3, 4)$ and $s = 3$.

THEOREM 7.21. *Let* $\mathbf{S}^\lambda$ *be a graded marked surface and* (S, ξ) *a Riemann surface with a CY-s type* q_s*-quadratic differential. If* $\mathfrak{N}(\mathbf{S}^\lambda) = \mathfrak{N}(\mathrm{S}, \xi)$*, then there is a homeomorphism*

$$h\colon \log_{\mathbf{c}} \mathbf{S}_\Delta^\lambda \to \log \mathrm{S}_\Delta^\xi,$$

which commutes with the action of the deck transformation group $\langle q \rangle$*. Moreover, the numerical data* $\mathfrak{N} \log \mathrm{S}_\Delta^\xi$ *matches* $\mathfrak{N}(\log_{\mathbf{c}} \mathbf{S}_\Delta^\lambda)$ *under* h*.*

PROOF. The data $(g, b, \mathbf{k})$ is trivial to check while the tricky part is to show that winding numbers (i.e., index l_i) matches. This becomes the calculation of winding numbers in the following two situations:

- of a loop γ around an exponential type singularity, where the local coordinate of the quadratic differential is in the form of $e^{z^{-k}} z^{-l} g(z) \, \mathrm{d} z^{\otimes 2}$ as in (6.7). This gives the grading change in $\log_{\mathbf{c}} \mathbf{S}_\Delta^\lambda$;
- of a loop γ that contains an s-pole of type (k, l) together with k s-simple zeroes. This gives the corresponding grading change in $\log \mathrm{S}_\Delta^\xi$, cf. left picture in Figure 8.1.

The two calculations are shown in Proposition 7.26 and Proposition 7.27, respectively, in the next subsection. As the Lekili-Polishchuk data LP_g is determined by the winding function, the theorem follows. □

7.5. Winding numbers

In this subsection, we compute angle changes/winding numbers of certain cycles on a Riemann surface with a quadratic differential.

First we recall the definition of the curvature for smooth paths in $\mathbb{R}^2$ with the classical standard flat metric $\mathrm{d} x^2 + \mathrm{d} y^2$. Take a coordinate $(x, y) \in \mathbb{R}^2$ and consider a smooth path

$$\gamma\colon [0, 1] \to \mathbb{R}^2, \quad t \mapsto \gamma(t) = (x(t), y(t)).$$

The *curvature* $\kappa(t)$ of a path $\gamma(t)$ is defined by

$$\kappa(t) := \frac{\dot{x}\ddot{y} - \ddot{x}\dot{y}}{(\dot{x}^2 + \dot{y}^2)^{\frac{3}{2}}},$$

where $\dot{x}(t) = \frac{\mathrm{d}}{\mathrm{d}\,t}x(t)$ and $\ddot{x}(t) = \frac{\mathrm{d}^2}{\mathrm{d}\,t^2}x(t)$. Then we define the *angle change of a path* γ by

$$\mathrm{Ang}(\gamma) := \int_0^1 \kappa(t)|\dot{\gamma}(t)|\,\mathrm{d}\,t = \int_0^1 \frac{\dot{x}\ddot{y} - \ddot{x}\dot{y}}{\dot{x}^2 + \dot{y}^2} dt.$$

If we identify $\mathbb{R}^2$ with $\mathbb{C}$ by $w = x + \mathbf{i}y$, then the angle change can be written as

$$\mathrm{Ang}(\gamma) = \mathrm{Im} \int_0^1 \frac{\overline{\dot{w}}\ddot{w}}{|\dot{w}|^2}\,\mathrm{d}\,t = \mathrm{Im} \int_0^1 \frac{\ddot{w}}{\dot{w}}\,\mathrm{d}\,t.$$

We note that if two paths γ and γ' are regularly homotopic and have the same tangent vectors at boundaries $t = 0, 1$, then $\mathrm{Ang}(\gamma) = \mathrm{Ang}(\gamma')$. We also note that by definition, the angle change is invariant under the transformation $w \mapsto \alpha w + \beta$, where $\alpha \in \mathbb{C}^*$ and $\beta \in \mathbb{C}$ are some constants.

Now we extend the definition of the angle change on a Riemann surface S with a quadratic differential ξ (of various type). Take a small open subset $U \subset \mathrm{S} \setminus \mathrm{Crit}(\xi)$ with a coordinate $z \in U$ and consider the distinguished coordinate

$$w = \int \sqrt{\xi}.$$

For a smooth path

$$\gamma\colon [0,1] \to U, \quad t \mapsto \gamma(t) = z(t),$$

we can similarly define the angle change of γ by

$$\mathrm{Ang}(\gamma) := \mathrm{Im} \int_0^1 \frac{\ddot{w}(z(t))}{\dot{w}(z(t))} dt,$$

where $\dot{w}(z(t)) = \frac{\mathrm{d}}{\mathrm{d}\,t}w(z(t))$ and $\ddot{w}(z(t)) = \frac{\mathrm{d}^2}{\mathrm{d}\,t^2}w(z(t))$. In the q_s-quadratic differential setting, since the coordinate w is determined up to

$$w \mapsto \pm q_s^m w + c,$$

the angle change $\mathrm{Ang}(\gamma)$ is also well-defined. For a general path $\gamma\colon [0,1] \to \mathrm{S} \setminus \mathrm{Crit}(\xi)$, we can define the angle change by taking an open covering $\gamma \subset \cup_\alpha U_\alpha$ and using the distinguished coordinate w on each U_α.

LEMMA 7.22. *Assume that on an open subset $U \subset \mathrm{S} \setminus \mathrm{Crit}(\xi)$ with a coordinate $z \in U$, the quadratic differential ξ takes the form $\xi = f(z)\,\mathrm{d}\,z^{\otimes 2}$. Then for a path $\gamma\colon [0,1] \to U$ with the z-coordinate expression $z(t)$, we have*

$$\mathrm{Ang}(\gamma) = \frac{1}{2}\,\mathrm{Im} \int_0^1 \frac{f'(z(t))}{f(z(t))}\dot{z}(t)dt + \mathrm{Im} \int_0^1 \frac{\ddot{z}(t)}{\dot{z}(t)}dt,$$

where $f'(z) = \frac{\mathrm{d}}{\mathrm{d}\,t}f(z)$.

PROOF. First we note that by definition $\frac{\mathrm{d}}{\mathrm{d}\,t}w(z) = \sqrt{f(z)}$. So we have

$$\frac{\mathrm{d}}{\mathrm{d}\,t}w(z(t)) = \sqrt{f(z(t))}\dot{z}(t),$$

$$\frac{\mathrm{d}^2}{\mathrm{d}\,t^2}w(z(t)) = \frac{f'(z(t))}{2\sqrt{f(z(t))}}\dot{z}(t)^2 + \sqrt{f(z(t))}\ddot{z}(t).$$

Then by the definition of $\mathrm{Ang}(\gamma)$, we have the following alternative definition of winding numbers. □

DEFINITION 7.23. When γ is a loop, the *winding number* is given by (cf. Definition 3.2)

$$\mathrm{wind}_\xi(\gamma) := \frac{1}{\pi}\mathrm{Ang}(\gamma). \tag{7.18}$$

We now compute the winding numbers around s-simple zeroes, s-poles (of type (k,l)) and exponential type singularities.

LEMMA 7.24. *Consider a q_s-quadratic differential $\xi = z^{s-2}\,\mathrm{d}\,z^{\otimes 2}$ on $\mathbb{C}$. Then for a small loop γ around s-simple zero at $0 \in \mathbb{C}$, we have*

$$\mathrm{wind}_\xi(\gamma) = \mathrm{Re}(s). \tag{7.19}$$

PROOF. Set $z(t) = e^{\mathbf{i}t}$. Then by Lemma 7.22, we have

$$\begin{aligned}
\mathrm{Ang}(\gamma) &= \frac{1}{2}\,\mathrm{Im}\int_0^{2\pi}\frac{(s-2)z(t)^{s-3}}{z(t)^{s-2}}\dot{z}(t)\,\mathrm{d}\,t + \mathrm{Im}\int_0^{2\pi}\frac{\ddot{z}(t)}{\dot{z}(t)}dt\\
&= \frac{1}{2}\,\mathrm{Im}\int_0^{2\pi}\frac{(s-2)e^{\mathbf{i}t(s-3)}}{e^{\mathbf{i}t(s-2)}}\mathbf{i}e^{\mathbf{i}t}\,\mathrm{d}\,t + \mathrm{Im}\int_0^{2\pi}\frac{\mathbf{i}^2e^{\mathbf{i}t}}{\mathbf{i}e^{\mathbf{i}t}}dt\\
&= \frac{1}{2}\,\mathrm{Im}\,2\pi\mathbf{i}(s-2) + \mathrm{Im}\,2\pi\mathbf{i}\\
&= \pi\,\mathrm{Re}(s).
\end{aligned}$$

□

Similarly, we have the following.

LEMMA 7.25. *Consider a q_s-quadratic differential $\xi = z^{-k(s-2)-l}\,\mathrm{d}\,z^{\otimes 2}$ on $\mathbb{C}$. Then for a small loop γ around s-pole of type* $(\mathbf{k}.\mathbf{l})$ *at $0 \in \mathbb{C}$, we have*

$$\mathrm{wind}_\xi(\gamma) = -k(\mathrm{Re}(s)-2) - l + 2. \tag{7.20}$$

PROPOSITION 7.26. *Let ξ be a q_s-quadratic differential on* S *and $U \subset$ S be a simply connected domain which contains just k s-simple zeroes and an s-pole of type (k,l). Then for a loop $\gamma \subset U$ encircling these zeroes and a pole, we have*

$$\mathrm{wind}_\xi(\gamma) = 2 - l. \tag{7.21}$$

PROOF. This follows from combing (7.19), (7.20) and the following fact:

- if the loop γ is obtained from a loop γ' by including a single s-simple zero, then

$$\mathrm{wind}_\xi(\gamma) = \mathrm{wind}_\xi(\gamma') + \mathrm{Re}(s-2).$$

The proof of this claim is illustrated in Figure 7.1. □

Set $D = \{\, z \in \mathbb{C} \mid |z| < 2\,\}$.

PROPOSITION 7.27. *Consider a quadratic differential $\phi = e^{z^{-k}}z^{-l}g(z)\,\mathrm{d}\,z^{\otimes 2}$ on D with an exponential type singularity of index (k,l) at $0 \in D$, where $g(z)$ is non-zero holomorphic function on D. Then, for a small loop γ around* 0, *we have*

$$\mathrm{wind}_\xi(\gamma) = 2 - l. \tag{7.22}$$

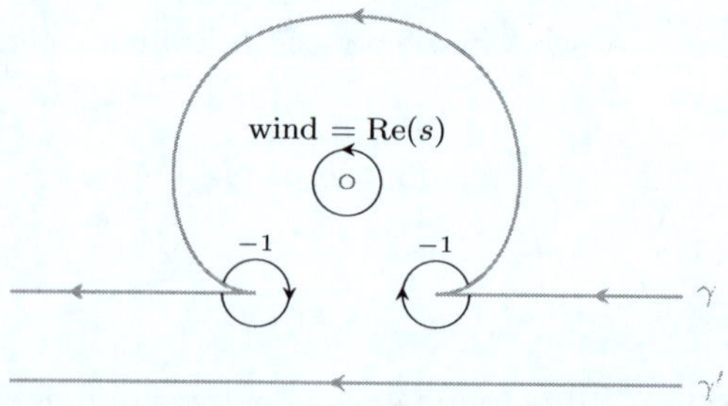

FIGURE 7.1. Winding changes Re(s) − 2 when goes around an s-simple zero

PROOF. Set $z(t) = e^{\mathbf{i}t}$. We note that the formula in Lemma 7.22 can be written as

$$\mathrm{Ang}(\gamma) = \frac{1}{2}\,\mathrm{Im}\int_\gamma \frac{\mathrm{d}}{\mathrm{d}\,z}\log f(z)\,\mathrm{d}\,z + \mathrm{Im}\int_0^1 \frac{\ddot{z}(t)}{\dot{z}(t)}dt.$$

Then the first term is

$$\begin{aligned}
&\frac{1}{2}\,\mathrm{Im}\int_\gamma \frac{\mathrm{d}}{\mathrm{d}\,z}\log e^{z^{-k}} z^{-l} g(z)\,\mathrm{d}\,z\\
=&\frac{1}{2}\,\mathrm{Im}\int_\gamma \frac{\mathrm{d}}{\mathrm{d}\,z}\left(z^{-k} - l\log z + \log g(z)\right)\mathrm{d}\,z\\
=&\frac{1}{2}\,\mathrm{Im}\int_\gamma \left(-kz^{-k-1} - lz^{-1} + \frac{g'(z)}{g(z)}\right)\mathrm{d}\,z\\
=&\frac{1}{2}\,\mathrm{Im}(-2\pi\mathbf{i}l)\\
=&-\pi l.
\end{aligned}$$

Here we use the two facts:

- the residue of z^{-k-1} at $z = 0$ is zero since $k \geq 1$ and
- the residue of $g'(z)/g(z)$ at $z = 0$ is zero since this function is holomorphic near zero by the condition that $g(z)$ is non-vanishing on D.

Next, we can easily compute that the contribution of the second term is 2π. Thus we have the proposition follows. □

7.6. Surface framing and period maps

Let $\mathbf{S}^\lambda$ be a graded marked surface with numerical data $\mathfrak{N}(\mathbf{S}^\lambda)$ and an associated graded DMS $(\mathbf{S}_\Delta, \mathbf{c}, \Lambda)$. Let $\log_{\mathbf{c}} \mathbf{S}_\Delta^\lambda$ be its topological log surface (cf. Definition 5.10). Fix $s \in \mathbb{C}$.

DEFINITION 7.28. A $\log_{\mathbf{c}} \mathbf{S}_\Delta^\lambda$*-framed q-quadratic differential* $\Theta = (\mathrm{S}, \xi, h; s)$ consists of a CY-s type q_s-quadratic differential ξ on a genus g Riemann surface S of s-polar type $(\mathbf{k}, \mathbf{l})$ together with an isotopy class of a homeomorphism

$$h\colon \log_{\mathbf{c}} \mathbf{S}_\Delta^\lambda \to \log \mathrm{S}_\Delta^\xi, \tag{7.23}$$

such that

- h identifies the decoration Δ of $\mathbf{S}_\Delta$ with the set Zero(ξ) of zeroes of ξ,
- h sends $\partial \log_{\mathbf{c}} \mathbf{S}_\Delta^\lambda$ to the set of real blow-ups of poles Pol(ξ) of ξ,

- h commutes with the action of the deck transformation group $\langle q\rangle$ and
- h matches the numerical data $\mathfrak{N}(\log \mathrm{S}_\Delta^\xi)$ with $\mathfrak{N}(\log_{\mathbf{c}} \mathbf{S}_\Delta^\lambda)$.

Two q-quadratic differential $\Theta_i = (\mathrm{S}_k, \xi_i, h_i; s)$ are $\log_{\mathbf{c}} \mathbf{S}_\Delta^\lambda$-equivalent if

- there is a biholomorphism $F\colon \mathrm{S}_1 \to \mathrm{S}_2$ satisfying $F(\mathrm{Pol}(\xi_1)) = \mathrm{Pol}(\xi_2)$, together with the choice of a lift $\widetilde{F}\colon \log \mathrm{S}_1 \to \log \mathrm{S}_2$ such that $\widetilde{F}^*\xi_2 = \xi_1$ and
- $h_2^{-1} \circ \log \widetilde{F} \circ h_1 \in \mathrm{Homeo}_0(\log_{\mathbf{c}} \mathbf{S}_\Delta^\lambda)$, where

$$\log \widetilde{F}\colon (\log \mathrm{S}_{1,\Delta}^\circ)^{\xi_1} \to (\log \mathrm{S}_{2,\Delta}^\circ)^{\xi_2}$$

  is the induced homeomorphism.

Here, $\mathrm{Homeo}_0(\log_{\mathbf{c}} \mathbf{S}_\Delta^\lambda)$ consists of isotopy classes of homeomorphisms of $\log_{\mathbf{c}} \mathbf{S}_\Delta^\lambda$ that commute with the deck transformation and are isotopy to identity.

Denote by $\mathrm{QQuad}_s(\log_{\mathbf{c}} \mathbf{S}_\Delta^\lambda)$ the moduli space of $\log_{\mathbf{c}} \mathbf{S}_\Delta^\lambda$-framed q-quadratic differential q-quadratic differentials.

Consider the spectral cover $\log \widehat{\mathrm{S}}^\circ \to \log \mathrm{S}^\phi$. This can be naturally extended to the double cover $(\log \widehat{\mathrm{S}}_\Delta^\circ)^\xi \to (\log \mathrm{S}_\Delta^\phi)^\xi$ with metric completions at zeroes and oriented blow-ups at poles. Correspondingly, we can also define the spectral cover of the topological log surface $\log \widehat{\mathbf{S}}_\Delta \to \log_{\mathbf{c}} \mathbf{S}_\Delta^\lambda$ with the covering involution τ^Δ such that $h\colon \log_{\mathbf{c}} \mathbf{S}_\Delta^\lambda \to \log \mathrm{S}_\Delta^\xi$ lifts to the homeomorphism between spectral covers $\widehat{h}\colon \log \widehat{\mathbf{S}}_\Delta \to (\log \widehat{\mathrm{S}}_\Delta^\circ)^\xi$. Finally, the framing h in (7.23) induces an isomorphism of the hat homology groups

$$\widehat{h}_*\colon \widehat{\mathbf{H}}(\log_{\mathbf{c}} \mathbf{S}_\Delta^\lambda) \to \widehat{\mathbf{H}}(\xi),$$

where $\widehat{\mathbf{H}}(\log_{\mathbf{c}} \mathbf{S}_\Delta^\lambda)$ is the corresponding hat homology group (i.e. defined by $\tau_*^\Delta\gamma = -\gamma$).

Recall that the period (7.15) of ξ is an R-linear map. Thus a $\log_{\mathbf{c}} \mathbf{S}_\Delta^\lambda$-framed q_s-quadratic differential (ξ, h) gives an R-linear map

$$Z_\xi \circ \widehat{h}_* \in \mathrm{Hom}_R(\widehat{\mathbf{H}}(\log_{\mathbf{c}} \mathbf{S}_\Delta^\lambda), \mathbb{C}_s).$$

DEFINITION 7.29. The *period map*

$$\Pi_s\colon \mathrm{QQuad}_s(\log_{\mathbf{c}} \mathbf{S}_\Delta^\lambda) \to \mathrm{Hom}_R(\widehat{\mathbf{H}}(\log_{\mathbf{c}} \mathbf{S}_\Delta^\lambda), \mathbb{C}_s)$$

is defined by sending $(\xi, h) \mapsto Z_\xi \circ \widehat{h}_*$.

CONJECTURE 7.30. *The period map* Π_s *is a local homeomorphism.*

In the next section, we interpret q-quadratic differentials as q-stability conditions. Along the way, we will prove the conjecture above under certain assumptions.

7.7. Quadratic differentials on log surface

REMARK 7.31. Note that the monodromy representation ρ_s defined in Section 7.1, cf. (7.8), factors through the representation ρ above. More precisely, define a group homomorphism

$$\begin{array}{rccc} \mathrm{ev}_s^{\otimes 2}\colon & \langle q\rangle & \to & \mathbb{C}^* \\ & q & \mapsto & q_s^2, \end{array}$$

then we have $\rho_s = \mathrm{ev}_s^{\otimes 2} \circ \rho$. This implies that the CY-s type q_s-quadratic differential ξ is a single-valued non-zero holomorphic quadratic differential on $\log \mathrm{S}^\phi$ satisfying

$$q^*\xi = q_s^2 \xi. \tag{7.24}$$

Therefore, we have the following alternative definition of $\log_{\mathbf{c}} \mathbf{S}_\Delta^\lambda$-framed q-quadratic differentials.

LEMMA 7.32. *A* $\log_{\mathbf{c}} \mathbf{S}_\Delta^\lambda$-framed q-quadratic differential $\Theta = (\mathrm{S}, \xi, h; s)$ *is equivalent to the data* $\Theta = (\log \mathrm{S}, \xi, h; s)$ *as follows:*

- *a genus g Riemann surface* $\log \mathrm{S}$ *with an automorphism q.*
- *a meromorphic quadratic differential ξ on* $\log \mathrm{S}$ *such that*
  - *there are $\aleph = |\Delta|$ zeroes of ξ, each of which is a conical singularity of infinity angle, as in Definition* 6.9, *with local coordinate of the form*
  $$\xi = c e^{s\omega} \, \mathrm{d}\, \omega^{\otimes 2}.$$
  Denote by $\Delta = \{Z_j\}$ the set of these zeroes.
  - *there are $k_i \cdot b$ poles $\{p_i^j \mid 1 \le i \le b, j \in \mathbb{Z}_{k_i}\}$ of ξ such that the pole p_i^j is a conical singularity of $k_i\pi$ angle, with local coordinate of the form*
  $$\xi = c e^{-(k_i(s-2)+l_i-2)\omega} \, \mathrm{d}\, \omega^{\otimes 2}.$$
  - *The automorphisms q and ξ satisfy* (7.24). *Thus, q preserves zeroes and $q(p_i^j) = p_i^{j+1}$ in particular.*
- *a homeomorphism* $h\colon \log_{\mathbf{c}} \mathbf{S}_\Delta^\lambda \to \log \mathrm{S}_\Delta^\xi$ *such that*
  - *the numerical data of* $\log \mathrm{S}_\Delta^\xi$ *(similarly defined as in Remark* 7.20*) matches with the one of* $\log_{\mathbf{c}} \mathbf{S}_\Delta^\lambda$ *under h.*
  - *the deck transformation q on* $\log_{\mathbf{c}} \mathbf{S}_\Delta^\lambda$ *becomes the automorphism q on* $\log \mathrm{S}_\Delta^\xi$, *where* $\log \mathrm{S}_\Delta^\xi$ *is the real blow-up of* $\log \mathrm{S}$ *at all the poles.*

CONVENTION. *Let* $\Theta = (\log \mathrm{S}_\Delta, \xi, h; s)$ *be a* $\log_{\mathbf{c}} \mathbf{S}_\Delta^\lambda$*-framed q-quadratic differential. Using h, we will pull back various structures (i.e., foliation, metric, etc.) of* $\log \mathrm{S}_\Delta^\xi$ *to* $\log_{\mathbf{c}} \mathbf{S}_\Delta^\lambda$ *and omit* $(\log \mathrm{S}_\Delta, h)$ *in the following discussion when there is no confusion. Namely, we will simply write $\Theta = \xi$ and call it a q-quadratic differential on* $\log_{\mathbf{c}} \mathbf{S}_\Delta^\lambda$.

CHAPTER 8

q-Stability conditions via q-quadratic differentials

Recall that we have the following topological setting:

- $\mathbf{S}^\lambda = (\mathbf{S}, \mathbf{M}, \mathbf{Y}, \lambda)$ is a graded marked surface,
- $\mathbf{A}$ is an (initial) full formal arc system of $\mathbf{S}$,
- $\mathbf{c}$ is the cut that is compatible with $\mathbf{A}$,
- $(\mathbf{S}_\Delta, \mathbf{c}, \Lambda)$ is a graded DMS associated to $\mathbf{S}^\lambda$.
- $\log_{\mathbf{c}} \mathbf{S}_\Delta^\lambda = \bigcup_{m\in\mathbb{Z}} \mathbf{S}_\Delta^{(m)}$ is the topological log surface of $\mathbf{S}_\Delta$ with grading $\log\lambda$, with respect to $\mathbf{c}$;

and categorical setting:

- $\mathcal{D}_\infty(\mathbf{A})$ is the topological Fukaya category associated to $\mathbf{S}$,
- $\mathcal{D}_\mathbb{X}(\mathbf{A})$ is the Calabi-Yau-$\mathbb{X}$ category of $\mathcal{D}_\infty(\mathbf{A})$,
- we identify $\mathcal{D}_\infty(\mathbf{A})$ with an $\mathbb{X}$-baric heart of $\mathcal{D}_\mathbb{X}(\mathbf{A})$ under (5.4).

Note that the cut $\mathbf{c}$ induces an isomorphism

$$\widehat{\mathbf{H}}_1(\log_{\mathbf{c}} \mathbf{S}_\Delta^\lambda) \cong \mathbf{H}_1(\mathbf{S}, \mathbf{M}; \mathbb{Z}_{\mathrm{sp}}) \otimes R. \tag{8.1}$$

8.1. q-Quadratic differentials with cuts

Denote by $\mathrm{Core}(\xi)$ the core of ξ on $\log_{\mathbf{c}} \mathbf{S}_\Delta^\lambda$. Let $\pi_\Delta \colon \log_{\mathbf{c}} \mathbf{S}_\Delta^\lambda \to \mathbf{S}_\Delta$ be the projection. Denote by

$$\underline{\mathrm{Core}}(\xi) \colon = \pi_\Delta(\mathrm{Core}(\xi)) \subset \mathbf{S}_\Delta$$

the projection of $\mathrm{Core}(\xi)$ on $\mathbf{S}_\Delta$. Note that $\underline{\mathrm{Core}}(\xi)$ is well-defined topologically since (7.24) implies that $\mathrm{Core}(\xi)$ and angles/direction of geodesics are invariants under the deck transformation (but the length of geodesics may be scaled). In order to give the metric information on $\underline{\mathrm{Core}}(\xi)$, we need to choose a cut. We will discuss the existence of cuts in the next subsection.

Definition 8.1. Let ξ be a q-quadratic differential on $\log_{\mathbf{c}} \mathbf{S}_\Delta^\lambda$. A cut $\mathbf{t} = \{t_i\}$ of $\mathbf{S}_\Delta$ is compatible with ξ if any of its arc t_i does not intersect the interior of $\underline{\mathrm{Core}}(\xi)$. Denote by $\mathrm{QQuad}_s^{\mathrm{c}}(\log_{\mathbf{c}} \mathbf{S}_\Delta^\lambda)$ the q-quadratic differentials on $\log_{\mathbf{c}} \mathbf{S}_\Delta^\lambda$ that admit some compatible cut.

Now take a q-quadratic differential ξ with compatible cut $\mathbf{t}$. Then by Lemma 5.9 there is a unique element $b_* \in \mathrm{SBr}(\mathbf{S}_\Delta)$ such that $b(\mathbf{t}) = \mathbf{c}$. Consider the induced action of b on $\log_{\mathbf{c}} \mathbf{S}_\Delta^\lambda$, which is also denoted by b. Denote by

$$\mathrm{Core}^0(\xi) \colon = \mathrm{Core}(b_*(\xi)) \cap \mathbf{S}_\Delta^{(0)} \tag{8.2}$$

the induced core on $\mathbf{S}_\Delta$, which is the intersection of the core $\mathrm{Core}(b_*(\xi))$ with the zero sheet $\mathbf{S}_\Delta^{(0)}$ of $\log_{\mathbf{c}} \mathbf{S}_\Delta^\lambda$. Note that ξ also provides foliation/metric on $\mathrm{Core}^0(\xi)$. By pulling the set of vertices Δ of $\mathrm{Core}^0(\xi)$ on $\mathbf{S}_\Delta$ along the arcs in $\mathbf{c}$ to the set $\mathbf{Y}$ of closed marked points of $\mathbf{S}_\Delta$, we obtain some convex hull $\mathrm{Core}^0_{\mathbf{c}}(\xi)$ (see Figure 8.1). More precisely, this is the inverse operation of $\mathbf{im_c}$ in (5.9):

$$\mathrm{Core}^0_{\mathbf{c}}(\xi) \colon = \mathbf{im}_{\mathbf{c}}^{-1}(\mathrm{Core}^0(\xi)) \subset \mathbf{S}^\lambda.$$

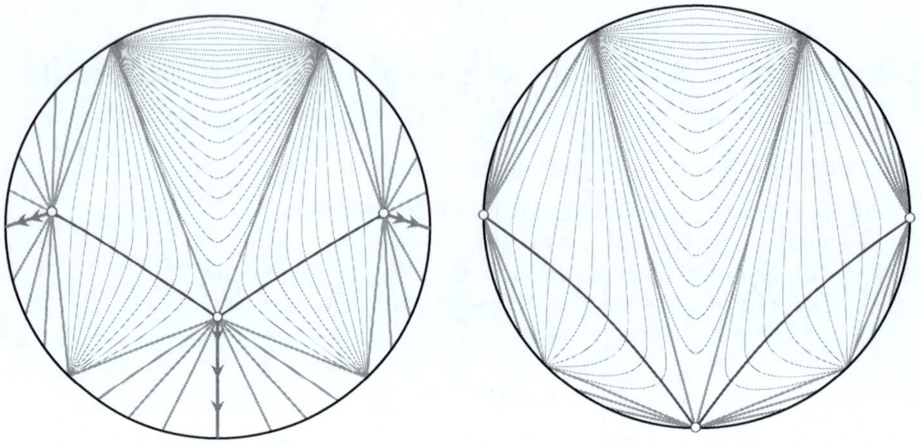

FIGURE 8.1. Horizonal strip decompositions on (a sheet of) $\log_{\mathbf{c}} \mathbf{S}^\lambda_\Delta$ and $\mathbf{S}$

The foliation of $\mathrm{Core}^0_{\mathbf{c}}(\xi)$ gives a partial foliation of $\mathbf{S}$, which can be extended to the rest of the surface by gluing upper half planes as the core determines a flat surface. Thus we obtain a flat surface, or equivalently, a quadratic differential ϕ on some graded marked surface $\mathbf{S}_0^{\lambda_0}$ determined by

$$\mathrm{Core}(\phi) = \mathrm{Core}^0_{\mathbf{c}}(\xi). \tag{8.3}$$

LEMMA 8.2. *ϕ is in* $\mathrm{FQuad}_\infty(\mathbf{S}^\lambda)$.

PROOF. By definition, we only need to show that the numerical data of ϕ (or of $\mathbf{S}_0^{\lambda_0}$) coincide with $\mathfrak{N}(\mathbf{S}^\lambda)$. This is exactly what we have shown in Theorem 7.21. □

By (6.11) in Theorem 6.13, ϕ corresponds to a stability condition

$$\widehat{\sigma}(\xi) \colon = \Xi_\infty(\phi) \in \mathrm{Stab}^\circ \mathcal{D}_\infty(\mathbf{A}). \tag{8.4}$$

Therefore, we obtain a triple $(\mathcal{D}_\infty(\mathbf{A}), \widehat{\sigma}(\xi), s)$. We proceed to show that such a triple induces a q-stability condition.

PROPOSITION 8.3. *Suppose that* $\mathrm{Re}(s) \geq 2$. *Let* $\xi \in \mathrm{QQuad}^{\mathrm{c}}_s(\log_{\mathbf{c}} \mathbf{S}^\lambda_\Delta)$. *Then the triple* $(\mathcal{D}_\infty(\mathbf{A}), \widehat{\sigma}(\xi), s)$, *constructed as above, induces a closed q-stability condition*

$$(\sigma, s) = (\mathcal{D}_\infty(\mathbf{A}), \widehat{\sigma}(\xi), s) \otimes_* R$$

on $\mathcal{D}_{\mathbb{X}}(\mathbf{A})$. *Then we obtain a map*

$$\Xi_s \colon \mathrm{QQuad}^{\mathrm{c}}(\log_{\mathbf{c}} \mathbf{S}^\lambda_\Delta) / \mathrm{SBr}(\mathbf{S}_\Delta) \to \mathrm{QStab}^* \mathcal{D}_{\mathbb{X}}(\mathbf{A}) / \mathrm{ST}, \tag{8.5}$$

which is in fact an isomorphism.

PROOF. Without loss of generality, assume b in (8.2) is trivial, i.e., ξ and $\mathbf{c}$ are compatible.

Step 1. We only need to prove (2.13), i.e., $\operatorname{gldim}\widehat{\sigma}+1\le\operatorname{Re}(s)$. Then Theorem 2.10 implies that the construction above gives a q-stability condition and we obtain a map Ξ_s. Suppose not, then there is a pair of semistable objects $\widehat{M_1}$ and $\widehat{M_2}$ in $\mathcal{D}_\infty(\mathbf{A})$ such that

$$\begin{cases}\operatorname{Hom}_{\mathcal{D}_\infty}(\widehat{M_1},\widehat{M_2})\neq 0,\\ \varphi_{\widehat{\sigma}}(\widehat{M_2})-\varphi_{\widehat{\sigma}}(\widehat{M_1})>\operatorname{Re}(s)-1(\geq 1),\end{cases}\tag{8.6}$$

where $\varphi_{\widehat{\sigma}}$ denotes the phase function with respect to $\widehat{\sigma}$.

By Theorem 3.7, $\widehat{M_i}$ correspond to graded closed arcs $\widetilde{\gamma}_i$ on $\mathbf{S}$, $i=1,2$, and the non-zero morphism in (8.6) corresponds to an intersection

$$p\in\widetilde{\gamma}_1\cap\widetilde{\gamma}_2$$

with $\operatorname{ind}_p(\widetilde{\gamma}_1,\widetilde{\gamma}_2)=0$. We claim that p can not be in the interior of $\mathbf{S}$ as shown below.

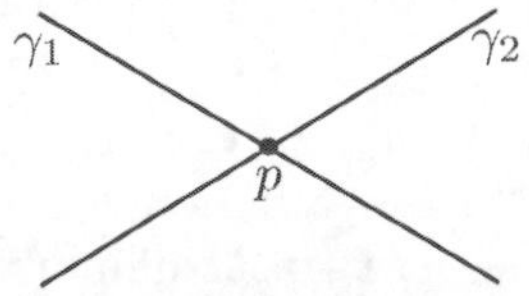

Otherwise, such an intersection in $\mathbf{S}^\circ$ between $\widetilde{\gamma}_i$ also contributes as

$$p\in\widetilde{\gamma}_2\cap\widetilde{\gamma}_1$$

with $\operatorname{ind}_p(\widetilde{\gamma}_2,\widetilde{\gamma}_1)=1$ by [**IQZ**, Lemma 2.3]. By Theorem 3.7 again, we also have a (non-zero) induced morphism in

$$\operatorname{Hom}_{\mathcal{D}_\infty}(\widehat{M_2},\widehat{M_1})[1].$$

As both $\widehat{M_i}$ (and their shifts) are (semi-)stable, we have

$$\varphi_{\widehat{\sigma}}(\widehat{M_1}[1])\geq\varphi_{\widehat{\sigma}}(\widehat{M_2})$$

which contradicts to the second inequality in (8.6).

Now let $Y\in\widetilde{\gamma}_1\cap\widetilde{\gamma}_2$ be the intersection, which is in $\mathbf{Y}$, corresponds to the non-zero morphism in (8.6). More precisely, such a morphism corresponds to the angle α from $\widetilde{\gamma}_1$ to $\widetilde{\gamma}_2$ at Y, that equals to (with respect to the quadratic differential ϕ)

$$\left(\varphi_{\widehat{\sigma}}(\widehat{M_2})-\varphi_{\widehat{\sigma}}(\widehat{M_1})\right)\cdot\pi.$$

Let $\widetilde{\eta}_i=\mathbf{im_c}(\widetilde{\gamma}_i)$, where $\mathbf{im_c}$ is defined in (5.9). Since $\widehat{M_i}$ are semistable, $\widetilde{\gamma}_i$ are in $\operatorname{Core}(\phi)=\operatorname{Core}^0_{\mathbf{c}}(\xi)$ and η_i are in $\operatorname{Core}^0(\xi)$. Denote by Z the corresponding intersection between η_i, which is the decoration connecting to Y by $\mathbf{c}$. By construction, the angle β from η_1 to η_2 at Z (with respect to the quadratic differential ξ) equals the angle α from $\widetilde{\gamma}_1$ to $\widetilde{\gamma}_2$ at Y, cf. Figure 8.2.

Let $\widetilde{\eta}_i$ be some lifts of η_i in $\log_{\mathbf{c}}\mathbf{S}^\lambda_\Delta$ so that β is also lifted to an angle $\widetilde{\beta}$ from $\widetilde{\eta}_1$ to $\widetilde{\eta}_2$. As the deck transformation q acts as rotation by $e^{\mathbf{i}\pi s}$ on q-quadratic differentials, the angle θ from η_2 to $q(\eta_1)$ at Z is

$$\begin{aligned}\theta&=\operatorname{Re}(s)\cdot\pi-\widetilde{\beta}\\&=\operatorname{Re}(s)\cdot\pi-\beta\\&=\operatorname{Re}(s)\cdot\pi-\alpha\\&=\operatorname{Re}(s)\cdot\pi-(\varphi(\widehat{M_2})-\varphi(\widehat{M_1}))\cdot\pi.\end{aligned}$$

By (8.6), $\theta < \pi$. Since both η_2 and $q(\eta_1)$ are in the core $\mathrm{Core}(\xi)$, which is a convex hull, there is a geodesic $\widetilde{\eta}_0$ in $\mathrm{Core}(\xi)$ obtained from $q(\eta_1) \cup \eta_2$ by smoothing the intersection at Z (on the side of the angle θ, cf. the violet dashed arc in Figure 8.2).

Then the projection $\pi_\Delta(\widetilde{\eta}_0)$ of $\widetilde{\eta}_0$ on $\mathbf{S}_\Delta$, which is in $\mathrm{Core}^0(\xi)$, will intersect the segment ZY in the cut $\mathbf{c}$. This contradicts to the compatibility between ξ and $\mathbf{c}$. Thus (2.13) holds and σ is a closed q-stability condition and we obtain a map Ξ_s in (8.5). The injectivity of Ξ_s follows from the facts that: Ξ_∞ in (6.11) is injective and the core determines quadratic differentials.

Step 2. Reversing the process above, we have the following.

Given a point in $\mathrm{QStab}^* \mathcal{D}_{\mathbb{X}}(\mathbf{A})/\mathrm{ST}$, it can be represented by a q-stability condition (σ, s) induced from some triple $(\mathcal{D}_\infty(\mathbf{A}), \widehat{\sigma}, s)$ for $\widehat{\sigma} \in \mathrm{Stab}^\circ \mathcal{D}_\infty(\mathbf{A})$. Let $\phi = \Xi_\infty^{-1}(\widehat{\sigma})$ be the exponential type quadratic differential on $\mathbf{S}^\lambda$ that corresponds to $\widehat{\sigma}$. Pushing the vertices of $\mathbf{Y}$ along the cut $\mathbf{c}$ to points in Δ, one gets a convex hull $\mathrm{Core}^0 = \mathbf{im_c}(\mathrm{Core}(\phi))$ on $\mathbf{S}_\Delta$ from the core $\mathrm{Core}(\phi)$. The convexity of Core^0 on $\log_{\mathbf{c}} \mathbf{S}_\Delta^\lambda$ follows from (2.13):

- Take two saddle connections, say $\widetilde{\eta}_i$ in Figure 8.2, that intersect at a decoration Z. Then the angle between them that does not intersect the cut, say from $\widetilde{\eta}_1$ to $\widetilde{\eta}_2$, is at most $\mathrm{gldim}(\widehat{\sigma}) \cdot \pi$. This implies the other angle, e.g. θ from $\widetilde{\eta}_2$ to $q(\widetilde{\eta}_1)$ in Figure 8.2, is at least

$$\mathrm{Re}(s) \cdot \pi - \mathrm{gldim}(\widehat{\sigma}) \cdot \pi \geq \pi.$$

Now we shall use this convex hull to build a q-quadratic differential on $\log_{\mathbf{c}} \mathbf{S}_\Delta^\lambda$. Consider the horizontal strip decomposition of $\mathbf{S}$ with respect to ϕ. Delete all the upper half planes that do not intersect the core $\mathrm{Core}(\phi)$. When pushing closed marked point in $\mathbf{Y}$ along the cut $\mathbf{c}$ to decorations in Δ, the foliation of the remaining of the horizontal strip decomposition provides a partial foliation/grading of the zero sheet $\mathbf{S}_\Delta^{(0)} \subset \log_{\mathbf{c}} \mathbf{S}_\Delta^\lambda$. Note that this partial foliation contains Core^0 mentioned above.

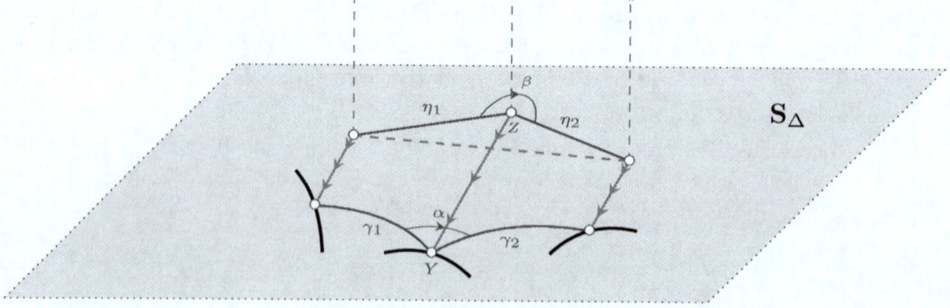

FIGURE 8.2. The cut (green) and lifts of closed arcs to log surface

The partial foliation for the m-th sheet $\mathbf{S}_\Delta^{(m)}$ is given in the same way from the horizonal strip decomposition on $\mathbf{S}$ with respect to the rotated quadratic differential $e^{mi\pi s} \cdot \phi$. Then it contains the corresponding convex hull Core^m obtained from $\mathrm{Core}(e^{mi\pi s} \cdot \phi)$.

The rest of the foliation on $\log_{\mathbf{c}} \mathbf{S}_\Delta^\lambda$ will be given/filled-in by part of the log-surface $C_\infty^s = s \log z$ around each decoration Z and by (non-essential) upper half planes. Here, $C_\infty^s = s \log z$ is topologically homeomorphic to C_∞ in (6.9). This can be done, i.e., the partial foliations do match because of the condition (2.13):

- Each angle of Core^m between saddle connections at a decoration Z is at most $\mathrm{gldim}\, \widehat{\sigma} \cdot \pi$. Therefore (the neighbourhood of) the cores Core^m around Z, do not overlap in C_∞^s.
- These foliations are compatible, i.e. they can be regarded as part of C_∞^s simultaneously, because by construction the deck transformation q becomes the rotation by $e^{mi\pi s}$.
- Thus we can fill in the gaps to get a foliation on $\log_{\mathbf{c}} \mathbf{S}_\Delta^\lambda$.

Finally, it is straightforward to check that the foliation we just constructed is indeed a q-quadratic differential on $\log_{\mathbf{c}} \mathbf{S}_\Delta^\lambda$ in Lemma 7.32. This completes the proof that Ξ_s is surjective. □

Now take a subspace $\mathrm{QQuad}_s^*(\log_{\mathbf{c}} \mathbf{S}_\Delta^\lambda)$ in $\mathrm{QQuad}_s^{\mathrm{c}}(\log_{\mathbf{c}} \mathbf{S}_\Delta^\lambda)$ which consists of those q-quadratic differentials, whose compatible cuts are in $\{b(\mathbf{c}) \mid b \in \mathrm{BT}(\mathbf{S}_\Delta)\}$. This is the analogue of taking a connected component $\mathrm{FQuad}_3^{\mathbb{T}}(\mathbf{S}_\Delta)$ in $\mathrm{FQuad}(\mathbf{S}_\Delta)$ in [**KQ2**, (4.13)], where they fit in the commutative diagram

$$(8.7)\qquad \begin{array}{ccc} \mathrm{FQuad}_3^{\mathbb{T}}(\mathbf{S}_\Delta) & \xrightarrow{\ \mathrm{BT}(\mathbf{S}_\Delta)\ } & \\ \big\downarrow & & \mathrm{FQuad}_3(\mathbf{S}) \\ \mathrm{FQuad}_3(\mathbf{S}) & \xrightarrow{\ \mathrm{SBr}(\mathbf{S})\ } & \end{array}$$

in that case (Calabi-Yau-3). As the subgroup $\mathrm{ST}\,\mathcal{D}_{\mathbb{X}}(\mathbf{A})$ of $\mathrm{SBr}(\mathbf{S}_\Delta)$ is isomorphic to $\mathrm{BT}(\mathbf{S}_\Delta)$ by Theorem 5.11, we have the following result, which is the Calabi-Yau-s version of Theorem 6.7, or the q-deformed version of Theorem 6.13.

THEOREM 8.4. *There is the following identification*

$$(8.8)\qquad \chi_s \colon \widehat{\mathbf{H}}(\log_{\mathbf{c}} \mathbf{S}_\Delta^\lambda) \to \mathrm{K}(\mathcal{D}_{\mathbb{X}}(\mathbf{A})).$$

Moreover, there is an isomorphism Ξ_s of complex manifolds that fits into the commutative diagram

$$(8.9)\qquad \begin{array}{ccc} \mathrm{QQuad}_s^{\mathrm{c}}(\log_{\mathbf{c}} \mathbf{S}_\Delta^\lambda) & \xrightarrow{\ \Xi_s\ } & \mathrm{QStab}_s^* \mathcal{D}_{\mathbb{X}}(\mathbf{A}) \\ \Big\downarrow{\scriptstyle \Pi_s} & & \Big\downarrow{\scriptstyle \mathcal{Z}_s} \\ \mathrm{Hom}_R(\widehat{\mathbf{H}}(\log_{\mathbf{c}} \mathbf{S}_\Delta^\lambda), \mathbb{C}_s) & \xrightarrow{\ \chi_*\ } & \mathrm{Hom}_R(\mathrm{K}(\mathcal{D}_{\mathbb{X}}(\mathbf{A})), \mathbb{C}_s), \end{array}$$

such that the period map Π_s is given by

$$\Pi_s = q_s \circ (\Pi_\infty \otimes 1) \colon \xi \mapsto \int \sqrt{\xi}$$

via (8.1) *and becomes the map* $\mathcal{Z}_s = q_s \circ (Z_\infty \otimes 1)$ *(i.e., central charge). More precisely,* (1.3) *holds for* $X = \widetilde{X}_{\mathbb{X}}$ *in* (5.7) *of Theorem* 5.11. *Here* Π_∞ *and* Z_∞ *are the ones in* (6.11).

Furthermore, for $(\sigma, s) = \Xi_s(\xi)$, *the map* $\widetilde{X}_{\mathbb{X}}$ *induces one-one correspondence between saddle connections of* ξ *on* $\log_{\mathbf{c}} \mathbf{S}_\Delta^\lambda$ *and* σ*-semistable objects in* $\mathcal{D}_{\mathbb{X}}(\mathbf{A})$.

PROOF. Firstly, (8.8) follows from (6.11) by tensoring R, noticing (2.6) and (8.1).

Secondly, (8.9) is the lifted version of (8.5) in Proposition 8.3, noticing that we have (5.8).

Finally, suppose that (σ, s) is induced from the triple $(\mathcal{D}_\infty(\mathbf{A}), \widehat{\sigma}, s)$ up to the action of $\operatorname{ST}\mathcal{D}_{\mathbb{X}}(\mathbf{A})$ and $\phi = \Xi_\infty^{-1}(\widehat{\sigma})$ as in the proof of Proposition 8.3. By Theorem 5.13, we deduce that the correspondence between σ-semistable objects and saddle connections of ξ also inherits from the correspondence between the $\widehat{\sigma}$-semistable objects and saddle connections of ϕ in Theorem 6.13. □

In particular, Conjecture 7.30 holds restricted to $\operatorname{QQuad}_s^*(\log_{\mathbf{c}} \mathbf{S}_\Delta^\lambda)$.

8.2. Existence of cuts

One technical issue is that if the compatible cuts always exist for a q-quadratic differential. In other words, such a existence is equivalent to the assumption below.

ASSUMPTION 8.5. $\operatorname{QQuad}_s^{\mathrm{c}}(\log_{\mathbf{c}} \mathbf{S}_\Delta^\lambda) = \operatorname{QQuad}_s(\log_{\mathbf{c}} \mathbf{S}_\Delta^\lambda)$.

We present some sufficient conditions for the existence of cuts in this subsection. First we have the following fact.

LEMMA 8.6. *The projection* $\pi_\Delta(Z)$ *of any zero/decoration* Z *of* ξ *on* $\mathbf{S}_\Delta$ *is in the boundary of* $\underline{\operatorname{Core}}(\xi)$.

PROOF. Up to rotation, we only need to consider the saddle-free case. Suppose that a zero Z is in the interior of $\underline{\operatorname{Core}}(\xi)$.

Let $K \gg 1$ be a positive integer. Consider all saddle connections starting from Z that are in sheets $\bigcup_{i=1}^K \mathbf{S}_\Delta^i$. Let l_i be the number of saddle connections in the sheet $\mathbf{S}_\Delta^i$. As mentioned above, ξ satisfies (7.24) and hence the deck transformation q preserves geodesics/saddle connections. Thus l_i are the same, denoted by l.

On the one hand, each of these saddle connections is a horizontal strip H, such that H is incident at Z with angle π. Thus the sum of angles at Z of horizontal strips that contains these saddle connections satisfies

$$K \cdot l \cdot \pi < K \cdot \operatorname{Re}(s) \cdot \pi + 2\pi,$$

where $\operatorname{Re}(s) \cdot \pi$ is the angle at Z for each sheet. Dividing by K of the inequality above and taking $\lim_{K \to +\infty}$ we see that $l \leq \operatorname{Re}(s)$.

On the other hand, the angle between any two adjacent saddle connections at Z is less than π, since $\operatorname{Core}(\xi)$ is convex and Z is in the interior of $\underline{\operatorname{Core}}(\xi)$. So the sum of angles at Z of horizontal strips that contains these saddle connections satisfies

$$2\pi + (K \cdot l - 1) \cdot \pi > K \cdot \operatorname{Re}(s) \cdot \pi.$$

Similarly, we deduce that $l \geq \operatorname{Re}(s)$, which forces $l = \operatorname{Re}(s)$. However, this will further force that the angle between any two adjacent saddle connection at Z equals π, which contradicts to the fact that Z is in the interior of $\underline{\operatorname{Core}}(\xi)$. □

Now we show that Assumption 8.5 always holds in the disk case.

COROLLARY 8.7. *If the genus g of* $\mathbf{S}$ *is zero, then Assumption* 8.5 *holds.*

PROOF. By Lemma 8.6, any decoration $Z \in \Delta$ is in the boundary of $\underline{\mathrm{Core}}(\xi)$, so that there is no obstruction to pair points in Δ and $\mathbf{Y} \in \partial \mathbf{S}_\Delta$. Hence the corollary holds. □

Also, when $\mathrm{Re}(s) \gg 1$, then Assumption 8.5 holds. More precisely, we have the following.

PROPOSITION 8.8. *Recall that* $(\mathbf{k}, \mathbf{l}) := \{(k_1, l_1), \ldots, (k_b, l_b)\}$. *If*

$$\mathrm{Re}(s) \geq \max_{1 \leq i \leq b} \{3 - k_i - l_i\}, \tag{8.10}$$

then Assumption 8.5 *holds.*

PROOF. We use the winding number and a combinatorial fact to prove the proposition.

Step 1. Since the boundaries of $\log_{\mathbf{c}} \mathbf{S}_\Delta^\lambda$ contract to the core $\mathrm{Core}(\xi)$, the boundaries of $\mathbf{S}_\Delta$ contract to $\underline{\mathrm{Core}}(\xi)$. If we cut $\mathbf{S}_\Delta$ along $\underline{\mathrm{Core}}(\xi)$, we will get exactly b annulus $\mathbb{A}_i$, each of which contains a boundary component ∂_i of $\mathbf{S}_\Delta$. Equivalently, we have

$$\mathbf{S}_\Delta \setminus \underline{\mathrm{Core}}(\xi) \cong \bigcup_{i=1}^{b} \mathbb{A}_i, \quad \partial_i \subset \partial \mathbb{A}_i,$$

where $\mathbb{A}_i$ can intersect only at boundaries:

$$\mathbb{A}_i^\circ \cap \mathbb{A}_j^\circ = \emptyset, \quad i \neq j.$$

Suppose that ∂_i is the real blow-up of a pole p_i of type (k_i, l_i) and $Z_1, \ldots, Z_k$ are the s-simple zeroes on the other boundary $C_i (\neq \partial_i)$ of $\mathbb{A}_i$.

Let L be the unique simple closed curve in $\mathbb{A}_i$ (up to isotopy), clockwise around C_i. The winding number of L, as an anticlockwise loop around ∂_i is

$$\mathrm{wind}(L) = -\mathrm{wind}_\xi(\partial_i) = k_i(\mathrm{Re}(s) - 2) + l_i - 2.$$

On the other hand, such a winding number can be calculated as a clockwise loop around C_i:

- boundary arcs (connecting zeroes) of C_i are (images of) saddle connections (with zero winding) and;
- around each s-simple zero Z_j, the winding number is $\mathrm{Re}(s)$, which implies that Z_j contributes the winding of L is
$$\mathrm{wind}(L; Z_j) = \alpha - 1 < \mathrm{Re}(s) - 1,$$
for $0 < \alpha < \mathrm{Re}(s)$ is the angle at the corner Z_j of C_j shown in Figure 8.3.

Thus, we have the estimation that

$$\mathrm{wind}(L) = \sum_{j=1}^{k} \mathrm{wind}(L; Z_j) < k \cdot (\mathrm{Re}(s) - 1)$$

or

$$k > k_i + \frac{k_i + l_i - 2}{\mathrm{Re}(s) - 1}. \tag{8.11}$$

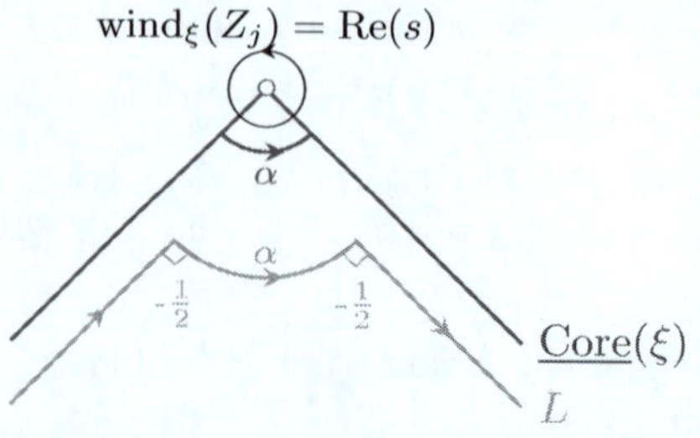

FIGURE 8.3. Winding contribution $1 + \alpha$ around a corner Z_j of C_j

By (8.10), we deduce that $k > k_i - 1$ or $k \geq k_i$.

Step 2. Consider a bipartite (unoriented) graph G, consisting of white vertices W_Z labelled by s-simple zeroes Z in Δ and black vertices B_{∂_i} labelled by boundary components ∂_i of $\mathbf{S}_\Delta$. There is an edge in G connecting the white vertex W_Z and the black vertex B_{∂_i} if and only if $Z \in \Delta$ are in the boundary of $\mathbb{A}_i$. We associate a number k_i to B_{∂_i}, where (k_i, l_i) is the type of the corresponding s-pole of B_{∂_i}. We have the following conditions on G:

- **Step I** says that the valence of B_{∂_i} is at least k_i;
- Lemma 8.6 says that the valence of any W_Z is at least 1.

We claim that for any bipartite graph G with the conditions above, there is a matching E of G, consisting of $\aleph = \sum_{i=1}^{b} k_i$ edges of G, such that each white vertex W_Z is incident in exactly one edge of E and each black vertex B_{∂_i} is incident in exactly k_i edges of E.

Use induction on the number of edges in G, where the start step is trivial. For the inductive step, there are two cases:

- If there is a white vertex Z with valence 1, let B_{∂_i} be the only black vertex connected directly to Z (by an edge e). Then delete Z and reduce the associated number k_i of B_{∂_i} to $k_i - 1$ to get a graph G' which also satisfies the conditions above. By inductive assumption, there is a matching E' for G' and then $E' \cup \{e\}$ is a matching for G.
- Otherwise, the valence of each white vertex Z is at least 2. The sum of valences of all white vertices is at least
$$2\aleph = 2\sum_{i=1}^{b} k_i,$$
which is also the sum of valences of all black vertices. Then we deduce that the valence v_j of some black vertex B_{∂_i} is strictly bigger than the associated number k_i. Delete any edge incidents at B_j from G and we can apply the inductive assumption again.

Step 3. Combining the two facts above, the matching tells how to pair k_i s-simple zeroes to the boundary ∂_i, which then induces a cut (by connecting them in each of the annulus $\mathbb{A}_i$). In fact, the cut can be then chosen to be images of geodesics ('separating connections'). □

8.3. Conclusion

Up to the technical issue, i.e., Assumption 8.5 above, the spaces of induced open and closed q-stability conditions coincide, which provide the 'principal' components of the space $\mathrm{QStab}_s \mathcal{D}_{\mathbb{X}}(\mathbf{A})$ of q-stability conditions.

THEOREM 8.9. *Suppose that Assumption* 8.5 *holds (e.g. when* $\mathrm{Re}(s) \gg 1$*) and* $\mathrm{Re}(s) > 2$. *Then we have the following*

1°. $\mathrm{QStab}_s^{\oplus} \mathcal{D}_{\mathbb{X}}(\mathbf{A}) = \mathrm{QStab}_s^{*} \mathcal{D}_{\mathbb{X}}(\mathbf{A})$. *Thus,*

$$\mathrm{QQuad}_s(\log_{\mathbf{c}} \mathbf{S}_{\Delta}^{\lambda}) \cong \mathrm{QStab}_s^{\oplus} \mathcal{D}_{\mathbb{X}}(\mathbf{A}). \tag{8.12}$$

2°. *The image of the inclusion*

$$\mathrm{QStab}_s^{\oplus} \mathcal{D}_{\mathbb{X}}(\mathbf{A}) \hookrightarrow \mathrm{QStab}_s \mathcal{D}_{\mathbb{X}}(\mathbf{A})$$

is open and closed and hence consists of connected components.

We will write $\mathrm{QStab}_s^{\circ} \mathcal{D}_{\mathbb{X}}(\mathbf{A})$ *for* $\mathrm{QStab}_s^{\oplus} \mathcal{D}_{\mathbb{X}}(\mathbf{A})$ *in this case.*

PROOF. Take a closed q-stability condition (σ, s) in $\mathrm{QStab}_s^{*} \mathcal{D}_{\mathbb{X}}(\mathbf{A})$, which is induced from some triple, i.e.

$$(\sigma, s) = (\mathcal{D}_\infty(\mathbf{A}), \widehat{\sigma}, s) \otimes_* R$$

with $\mathrm{gldim}\, \widehat{\sigma} = \mathrm{Re}(s) - 1$ (here we take the canonical $\mathbb{X}$-baric heart without loss of generality). We will keep all the notations in Proposition 8.3, i.e.

- $\xi = \Xi_s^{-1}(\sigma, s)$ with a compatible cut $\mathbf{c}$ and the corresponding (exponential type) quadratic differential

$$\phi = \Xi_\infty^{-1}(\widehat{\sigma}) \in \mathrm{FQuad}_\infty(\mathbf{S}^\lambda),$$

cf. (8.3) and (8.4).

For the first statement, we need to show that there exists another compatible cut $\mathbf{c}_0$ such that the corresponding quadratic differential $\phi_0 \in \mathrm{FQuad}_\infty(\mathbf{S}^\lambda)$ and stability condition $\widehat{\sigma}_0 = \Xi_\infty(\phi_0)$ satisfying

$$(\sigma, s) = (\mathcal{L}_0(\mathcal{D}_\infty(\mathbf{A})), \widehat{\sigma}_0, s) \otimes_{\oplus} R \quad \text{with} \quad \mathrm{gldim}\, \widehat{\sigma}_0 < \mathrm{Re}(s) - 1.$$

We divide the proof of this statement into three steps.

Step 1: Consider a pair of semistable objects $\widehat{M_1}$ and $\widehat{M_2}$ in $\mathcal{D}_\infty(\mathbf{A})$ such that

$$\begin{cases} \mathrm{Hom}_{\mathcal{D}_\infty}(\widehat{M_1}, \widehat{M_2}) \neq 0, \\ \varphi_{\widehat{\sigma}}(\widehat{M_2}) - \varphi_{\widehat{\sigma}}(\widehat{M_1}) = \mathrm{gldim}\, \widehat{\sigma}. \end{cases} \tag{8.13}$$

Let $\widetilde{\gamma}_i$ be the closed arcs on $\mathbf{S}$ that correspond to $\widehat{M_i}$ in $\mathcal{D}_\infty(\mathbf{A})$. As in the proof of Proposition 8.3, they intersect at some $Y \in \mathbf{Y}$, such that the angle $\alpha = \mathrm{gldim}\, \widehat{\sigma} \cdot \pi$ at Y provides the non-zero morphism in (8.13), cf. Figure 8.2. By [**Q6**, Prop.5.7], there are only finitely many such angles α.

Step 2: Pulling along the cut $\mathbf{c}$, the arcs $\widetilde{\gamma}_i$ become $\widetilde{\eta}_i = \mathbf{im}_{\mathbf{c}}(\widetilde{\gamma}_i)$ on $\mathbf{S}_\Delta$ as before. Denote by $Z \in \Delta$ the decoration corresponding to Y and β the angle corresponding to α. So we have

$$\beta = \alpha = \mathrm{gldim}\, \widehat{\sigma} \cdot \pi = (\mathrm{Re}(s) - 1) \cdot \pi,$$

which implies the other angle θ at Z between $\widetilde{\eta}_2$ and $q(\widetilde{\eta}_1)$ is exactly π. This also means that $\widetilde{\eta}_i$ are boundary arcs of (the zero sheet of) the core $\mathrm{Core}^0(\xi)$.

Step 3: Now we can perturb the quadratic differential ξ a little bit to get a new one ξ_0, such that, the only changes happen at all those angles (β, θ) as in **Step 1**, namely, we require for each of such angles,

$$\beta_0 = \beta - \epsilon, \quad \theta_0 = \theta + \epsilon = \pi + \epsilon$$

for some small positive real number ϵ. By Assumption 8.5, ξ_0 admits a compatible cut $\mathbf{c}_0$, which is also compatible with ξ then.

Use the cut $\mathbf{c}_0$ to produce another quadratic differential $\phi_0 \in \mathrm{FQuad}_\infty(\mathbf{S}^\lambda)$ that corresponds to another stability condition $\widehat{\sigma}_0$ in $\mathrm{Stab}\,\mathcal{D}_\infty(\mathbf{A})$. Note that $\mathbf{c}_0$ will cut the angle β in **Step 2**. By [**Q6**, Prop.5.7],

$$\mathrm{gldim}\,\widehat{\sigma}_0 < \mathrm{gldim}\,\widehat{\sigma} = \mathrm{Re}(s) - 1.$$

With the Lagrangian immersion $\mathcal{L}_0$ that corresponds to $\mathbf{c}_0$, we obtain another triple to induce (σ, s):

$$(\sigma, s) = (\mathcal{L}_0(\mathcal{D}_\infty(\mathbf{A})), \widehat{\sigma}_0, s) \otimes_\oplus R,$$

as required.

For the second statement, we note that $\mathrm{QStab}_s^\oplus \mathcal{D}_\mathbb{X}(\mathbf{A})$ is open $\mathrm{QStab}_s \mathcal{D}_\mathbb{X}(\mathbf{A})$ since condition (2.12) is an open one (cf. [**IQ**, Thm. 5.11]). For the closedness in $\mathrm{QStab}_s \mathcal{D}_\mathbb{X}(\mathbf{A})$, one can use the analogue map K in [**BS**, § 11.5], as we have shown that these stability conditions are indeed coming from quadratic differentials. □

Part 3

Applications

CHAPTER 9

Bridgeland-Smith theory on Calabi-Yau-N categories

We keep all the settings at the beginning of Section 8.

9.1. N-reductions for categories

Suppose that $N \geq \operatorname{gldim} \mathcal{A}_{\mathbf{A}} + 1$ and $\Gamma^N_{\mathbf{A}}$ is the Calabi-Yau-N Ginzburg algebra obtained from $\Gamma^{\mathbb{X}}_{\mathbf{A}}$ by collapsing the double degree $a+b\mathbb{X}$ into a single degree $a+bN$. Then there is a projection between dg algebras

$$\pi_N \colon \Gamma^{\mathbb{X}}_{\mathbf{A}} \to \Gamma^N_{\mathbf{A}},$$

which induces a fully faithful exact functor

$$\pi_N \colon \mathcal{D}_{\mathbb{X}}(\mathbf{A})/[\mathbb{X}-N] \to \mathcal{D}_N(\mathbf{A}) \, (: \, = \mathcal{D}_{fd}(\Gamma^N_{\mathbf{A}})),$$

that makes $\mathcal{D}_N(\mathbf{A})$ the N-reduction of $\mathcal{D}_{\mathbb{X}} /\!\!/ [\mathbb{X}-N]$, in the sense that

- $\mathcal{D}_N(\mathbf{A})$ can be identified with a triangulated hull of $\mathcal{D}_{\mathbb{X}}(\mathbf{A})/[\mathbb{X}-N]$,
- $K(\mathcal{D}_N) \cong \mathbb{Z}^{\oplus n}$ and the induced R-linear map

  $$[\pi_N] \colon K(\mathcal{D}_{\mathbb{X}}(\mathbf{A})) \to K(\mathcal{D}_N(\mathbf{A}))$$

  is a surjection given by sending $q \mapsto (-1)^N$.

Thus, we have the following by [**IQ**, Thm. 4.5].

COROLLARY 9.1. *There is an embedding*

$$\operatorname{QStab}_N \mathcal{D}_{\mathbb{X}}(\mathbf{A}) \longrightarrow \operatorname{Stab} \mathcal{D}_N(\mathbf{A})$$

between complex manifolds. Denote by $\operatorname{Stab}^\circ \mathcal{D}_N(\mathbf{A})$ *the image of* $\operatorname{QStab}^*_N \mathcal{D}_{\mathbb{X}}(\mathbf{A})$.

We also give the following conjecture, which is the surface version of [**IQ**, Cor. 6.8].

CONJECTURE 9.2. *The cluster-$\mathbb{X}$ category $\mathcal{C}(\Gamma^{\mathbb{X}}_{\mathbf{A}})$ in (5.6) admits the Serre functor $\tau \circ [1]$ (which should correspond to $[\mathbb{X}-1]$), for τ the Auslander-Reiten functor. Moreover, the Calabi-Yau-$(N-1)$ cluster category $\mathcal{C}_{N-1}(\mathbf{A})$ of $\Gamma^N_{\mathbf{A}}$ is the (unique) triangulated hull of $\mathcal{C}(\Gamma^{\mathbb{X}}_{\mathbf{A}}) /\!\!/ \tau[2-N]$ that fits into the commutative diagram:*

$$\begin{array}{ccccccccc}
0 & \longrightarrow & \mathcal{D}_{fd}(\Gamma^{\mathbb{X}}_{\mathbf{A}}) & \longrightarrow & \operatorname{per}\Gamma^{\mathbb{X}}_{\mathbf{A}} & \longrightarrow & \mathcal{C}(\Gamma^{\mathbb{X}}_{\mathbf{A}}) & \longrightarrow & 0 \\
 & & \downarrow{\scriptstyle /\!\!/[\mathbb{X}-N]} & & \downarrow{\scriptstyle /\!\!/[\mathbb{X}-N]} & & \downarrow{\scriptstyle /\!\!/\tau[2-N]} & & \\
0 & \longrightarrow & \mathcal{D}_{fd}(\Gamma^N_{\mathbf{A}}) & \longrightarrow & \operatorname{per}\Gamma^N_{\mathbf{A}} & \longrightarrow & \mathcal{C}(\Gamma^N_{\mathbf{A}}) & \longrightarrow & 0.
\end{array} \tag{9.1}$$

REMARK 9.3. One interesting question is what is the condition for $\mathcal{D}_{\mathbb{X}}(\mathbf{A}) /\!\!/ [\mathbb{X}-N]$ to be N-reductive. The condition $N \geq \operatorname{gldim} \mathcal{A}_{\mathbf{A}} + 1$ here is actually for $\Gamma_{\mathbf{A}}^N$ being non-positive, so that it is a silting object in its perfect derived category and hence ensures $\operatorname{Stab} \mathcal{D}_N(\mathbf{A})$ is nonempty.

9.2. N-reductions for q-quadratic differentials

When $s = N \geq 3$ is a positive integer, then the q_s-quadratic differential ξ, locally of the form (7.4), becomes a classical type quadratic differential on some Riemann surface in Definition 6.3. We have the definition of generalized GMN differentials as follows.

DEFINITION 9.4. A *CY-N type differential* on a Riemann surface S is a meromorphic differential with classical type of singularities satisfying:

- all the zeroes have order $N-2$;
- the order of any pole is at least 3.

The (decorated) marked surface associated to (S, ϕ) is defined as in Definition 6.4.

Next, from a graded marked surface $\mathbf{S}^\lambda$ with associated graded DMS $\mathbf{S}_\Delta$, we construct the marked surface $\mathbf{S}_\Delta^N$ for q_N-quadratic differentials.

DEFINITION 9.5. The *(CY-N) marked surface* $\mathbf{S}^N$ is a marked surface modified from $\mathbf{S}^\lambda$ by

- forgetting the grading and closed marked points and
- changing the number of open marked points on the boundary ∂_i to be

$$k_i \cdot (N-2) + l_i - 2, \tag{9.2}$$

where (k_i, l_i) is the associated numerical data.

The *(CY-N) DMS* $\mathbf{S}_\Delta^N$ is modified from $\mathbf{S}_\Delta$ same as above.

Note that the condition (7.9) in Definition 7.5 that s-poles are higher s-poles ensures the number in (9.2) is a positive integer.

Similar to Definition 6.6, we have moduli spaces $\operatorname{FQuad}_N(\mathbf{S}_\Delta^N)$ of $\mathbf{S}_\Delta^N$-framed quadratic differentials. So we have the following straightforward observation, as the counterpart to Corollary 9.1.

LEMMA 9.6. *There is an embedding*

$$\operatorname{QQuad}_N(\log_{\mathbf{c}} \mathbf{S}_\Delta^\lambda) \longrightarrow \operatorname{FQuad}(\mathbf{S}_\Delta^N).$$

Denote its image by $\operatorname{FQuad}_N^\circ(\mathbf{S}_\Delta^N)$ *and the corresponding quotient space in* $\operatorname{FQuad}_N(\mathbf{S}^N)$ *by* $\operatorname{FQuad}_N^\circ(\mathbf{S}^N)$.

Combining Corollary 9.1 and Lemma 9.6 we have the following version of Theorem 8.9 as an N-analogue of Bridgeland-Smith's result (or King-Qiu's upgraded version). The condition $N \gg 1$ ensures that Assumption 8.5 holds.

THEOREM 9.7. *If $N \gg 1$, there is an isomorphism between complex manifolds*

$$\operatorname{FQuad}_N^\circ(\mathbf{S}_\Delta^N) \cong \operatorname{Stab}^\circ \mathcal{D}_{fd}(\Gamma_{\mathbf{A}}^N). \tag{9.3}$$

When $\mathbf{S}$ is the disk (i.e., type A case), the theorem above is the main theorem in [**I**]. However, notice that the method in [**I**] is a directly generalization of [**BS**] while our approach is different.

REMARK 9.8. Let us explain that in the unpunctured case of [**BS**], i.e., $N = 3$ and when the marked surface has no punctures (equivalent to the higher order pole condition), how the result above should reproduce (1.1). To see this, let $\mathbf{K} = (K_1, \cdots, K_b)$ be the numerical data of numbers of marked points on $\partial\mathbf{S}$ in the setting of [**BS**]. Recall that g and b are the genus and number of boundaries components of $\mathbf{S}^3$. One chooses a graded marked surface $\mathbf{S}^\lambda$ of genus g, with integers $\mathbf{l} = (l_1 \cdots, l_b)$ that sums to $4 - 4g$ and a set of integer $\mathbf{k} = (k_1 \cdots, k_b)$ for $k_i = K_i - l_i + 2$ as in (9.2). Here we need $k_i \geq 1$, which can be done since

- if $g \geq 1$ or $b \geq 2$, then
$$\sum_{i=1}^{b} k_i = \sum_{i=1}^{b} K_i + (4g - 4) + 2b \geq b;$$
- if $g = 0$ and $b = 1$, then $K_1 \geq 3$ (so that $\Delta \geq 1$ in (6.2)) and the estimation above also holds.

Then (9.3) in Theorem 9.7 indeed becomes (1.1) as required.

We hope to cover the punctured case of [**BS**, Theorem 1.2] in future works.

EXAMPLE 9.9. Let $\mathbf{S} = \mathbf{D}$ be a disk with 4 marked points on its boundary. So the numerical data is ($g = 0$, $b = 1$; $\mathbf{k} = (4)$, $\mathbf{l} = (4)$; $\mathrm{LP}_0 = \emptyset$).

Choose the full formal open arc system $\mathbf{A}$ with dual closed arc system $\mathbf{A}^*$ of $\mathbf{S}^\lambda$ as in the first picture in Figure 9.1. Note that the open arcs are drawn in blue

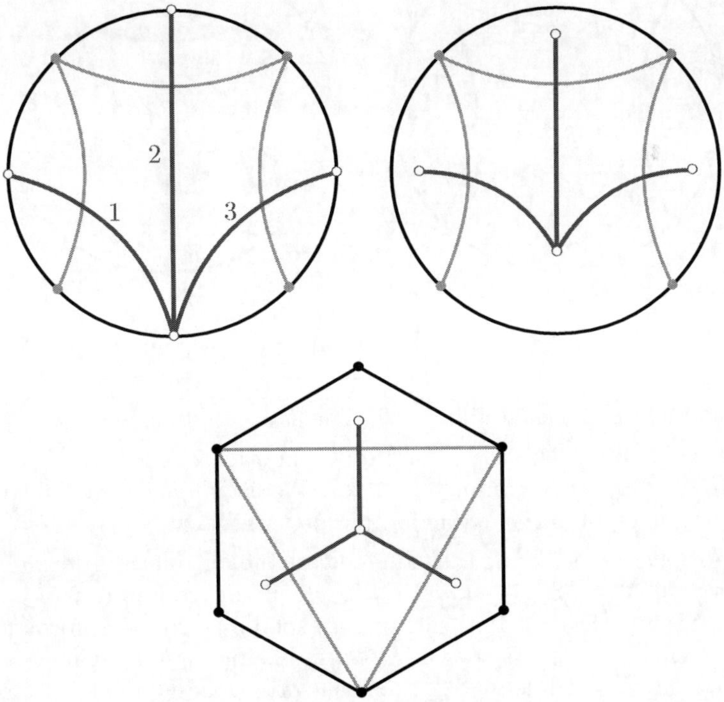

FIGURE 9.1. The A_3 case: $\mathbf{S}^\lambda$, $\mathbf{S}_\Delta$ and $\mathbf{S}^3_\Delta$ with arc systems

while the closed ones are in red. Suppose that the intersection indexes between open arcs in $\mathbf{A}$ are all zero. Then the corresponding dual closed arc systems $\mathbf{A}_\Delta^*$ in $\mathbf{S}_\Delta$ and $\mathbf{S}_\Delta^3$ are shown in the second and third pictures in Figure 9.1. Then we could have the corresponding graded quivers as:

(9.4)

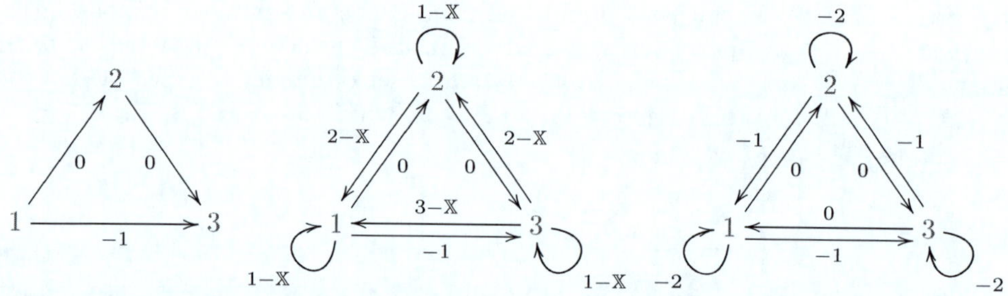

where the potentials associated to the second and third quivers are just the inner three cycles of degree $3 - \mathbb{X}$ and degree 0 respectively.

EXAMPLE 9.10. Keep the surface $\mathbf{S} = \mathbf{D}$ as above. Choose another full formal open arc system $\mathbf{A}$ with dual closed arc system $\mathbf{A}^*$ of $\mathbf{S}^\lambda$ as in the left picture in Figure 9.2. Then the corresponding dual full formal closed arc systems $\mathbf{A}_\Delta^*$ in $\mathbf{S}_\Delta$ is

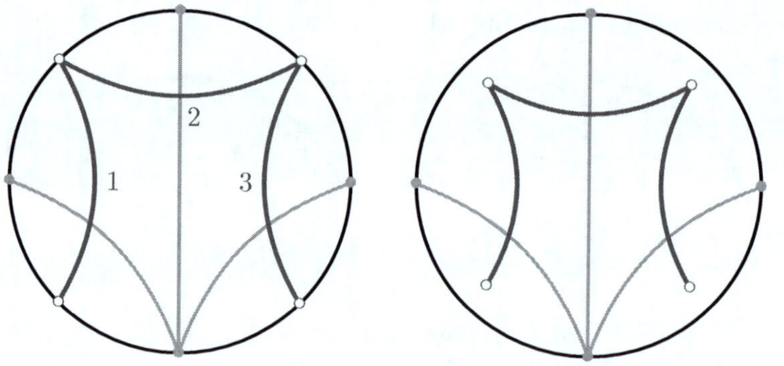

FIGURE 9.2. The A_3 case: alternative arc systems $\mathbf{S}^\lambda$ and $\mathbf{S}_\Delta$

shown in the right pictures in Figure 9.2. The associated quivers $(Q_\mathbf{A}, \widetilde{Q}_\mathbf{A})$ are just the standard A_3 quiver $1 \xleftarrow{a} 2 \xleftarrow{b} 3$ and its Calai-Yau-$\mathbb{X}$ double respectively, whose gradings are controlled by grading of curves. We list three cases of the gradings of $Q_\mathbf{A}$ with the corresponding 4-reduction pictures as follows:

- If $\deg a = \deg b = 0$, then the corresponding dual full formal closed arc system $\mathbf{A}_\Delta^*$ in $\mathbf{S}_\Delta^4$ is shown in the left picture of Figure 9.3.
- If $\deg a = \deg b = 1$, then the corresponding dual full formal closed arc system $\mathbf{A}_\Delta^*$ in $\mathbf{S}_\Delta^4$ is shown in the middle picture of Figure 9.3.
- If $\deg a = 1$ and $\deg b = 2$, then the corresponding dual full formal closed arc system $\mathbf{A}_\Delta^*$ in $\mathbf{S}_\Delta^4$ is shown in the right picture of Figure 9.3.

Note that the grading of quivers in the Calabi-Yau-N case is controlled by the numbers of marked points on the boundaries of $\mathbf{S}_\Delta^N$. In the final subsection of this section, we explain the cluster theory in the Calabi-Yau-N case.

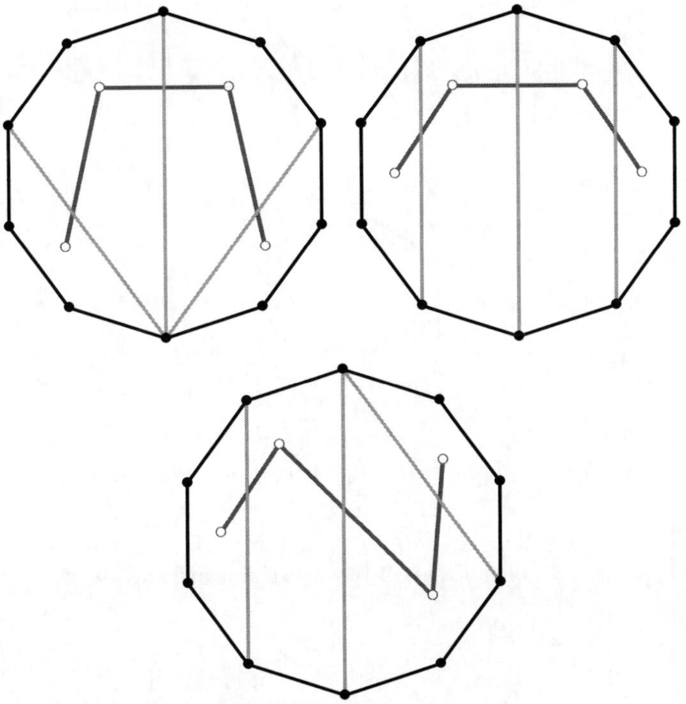

FIGURE 9.3. The A_3 case: three dual full formal arc systems in $\mathbf{S}^4_\Delta$

9.3. Cluster combinatorics

The skeletons (cf. [**KQ2**, Remark 2.17, Lemma 4.8 and Lemma 4.9], see also [**Q1**, **BS**]) for manifolds in (9.3) should be

$$\mathrm{EG}(\mathbf{S}^N) \cong \mathrm{CEG}(\mathcal{C}_{N-1}(\mathbf{A})), \tag{9.5}$$

where on the left hand side, it is the *exchange graph of N-angulation* of $\mathbf{S}_N$ and on the right hand side, it is the *cluster exchange graph* of $\mathcal{C}_{N-1}(\mathbf{A})$. The vertices of $\mathrm{EG}^N(\mathbf{S}^N)$ are N-angulation of $\mathbf{S}_N$ and the edges are forward flips. The vertices of $\mathrm{CEG}_{N-1}(\mathcal{C}(\mathbf{A}))$ are $(N-1)$-*cluster tilting sets* (cf. e.g. [**KQ1**, Def. 4.1]) in the cluster categories $\mathcal{C}_{N-1}(\mathbf{A})$ mentioned in Conjecture 9.2 with edges correspond to forward mutations.

For instance, Figure 9.4 shows a sequence of forward flips for $N = 4$-angulations of an 12-gon, which corresponds to meromorphic quadratic differentials of singularity type

$$(2^5, -14) = (2, 2, 2, 2, 2, -14)$$

on $\overline{\mathbb{C}} = \mathbb{P}^1$. More precisely, these 4-angulations consist of isotopy classes of separated saddle connections of the corresponding meromorphic quadratic differentials. Moreover, an example of a decorated forward flip for 4-angulation is from the middle picture of Figure 9.3 to the right picture there. For more details, cf. [**KQ2**, § 3].

One can also associate Calabi-Yau-N quivers with superpotential to N-angulations directly, parallel to Definition 5.1, where one replaces degree $\mathbb{X}$ with N. For

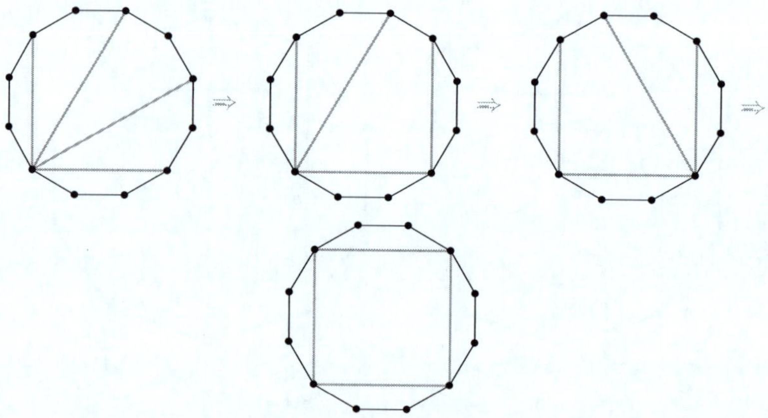

FIGURE 9.4. A sequence of flips of 4-angulations

instance, in figure 9.5 which is the case when $N = 5$, the superpotential should contains all those 3-cycle of degree -2 (two of them are shown in the figure).

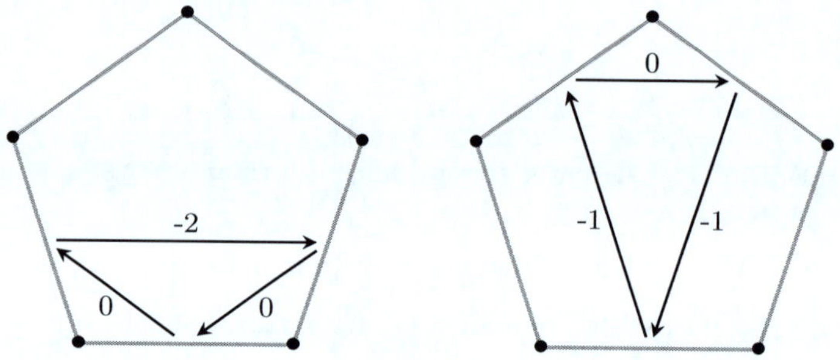

FIGURE 9.5. Some 3-cycles in the CY-5 case.

REMARK 9.11. Finally, we remark that one may want to get (9.3) by generalizing the approach in [**BS**], as [**I**] did for the type A case. However, this is not a short cut by all means (nor easy) due to lack of the following results:

- the topological cluster theory for surfaces, e.g.(9.5), has not been established.
- the categorical cluster theory for surfaces, e.g. [**KQ1**, Thm. 8.6] for $\Gamma_{\mathbf{A}}^{\mathbb{X}}$, has not been established.

CHAPTER 10

Hurwitz spaces via quadratic differentials in genus zero

The aim of this section is to identify the moduli space of CY-s type quadratic differentials with the Hurwitz space in the case $g = 0$.

10.1. Hurwitz spaces

DEFINITION 10.1. Let $\mathbf{k} = (k_1, \dots, k_b)$ be a multi-set of positive integers $k_i \in \mathbb{Z}_{\geq 1}$. A *Hurwitz cover of genus g and polar type* $\mathbf{k}$ is a pair of a compact Riemann surface S of genus g and a meromorphic function $f\colon \mathrm{S} \to \mathbb{P}^1$ such that poles of f consists of b points, i.e., $f^{-1}(\infty) = \{p_i\}_{i=1}^b$ with orders $\mathrm{ord}_f(p_i) = -k_i$.

Two Hurwitz covers (S, f) and (S', f') are equivalent if there is a biholomorphism $h\colon \mathrm{S} \to \mathrm{S}'$ such that $f = h^* f'$. The *Hurwitz space* $\mathrm{HS}(g, \mathbf{k})$ is the moduli space of Hurwitz covers of genus g and polar type $\mathbf{k}$.

For a Hurwitz cover (S, f), denote by $\mathrm{Zero}(f), \mathrm{Pol}(f) \subset S$ the set of zeroes and poles of f respectively. Write $\mathrm{Crit}(f) = \mathrm{Zero}(f) \cup \mathrm{Pol}(f)$. It is known that the dimension of the Hurwitz space is given by

$$\dim_{\mathbb{C}} \mathrm{HS}(g, \mathbf{k}) = 2g - 2 + b + \sum_{i=1}^{b} k_i,$$

which will be denoted by n.

The *regular locus* $\mathrm{HS}(g, \mathbf{k})_{\mathrm{reg}} \subset \mathrm{HS}(g, \mathbf{k})$ is defined to be the subset consisting of Hurwitz covers with simple zeroes, which will be called regular Hurwitz covers. For $(\mathrm{S}, f) \in \mathrm{HS}(g, \mathbf{k})_{\mathrm{reg}}$, denote $\Delta = \mathrm{Zero}(f)$. Then since

$$0 = \sum_{p \in \mathrm{Crit}(f)} \mathrm{ord}_f(p) = |\Delta| - \sum_{i=1}^{b} k_i,$$

we have $|\Delta| = \sum_{i=1}^{b} k_i$ in this case.

10.2. Marked surfaces and log surfaces from Hurwitz covers

In this subsection, we construct (ungraded) marked surfaces and (ungraded) log surfaces from Hurwitz covers.

CONSTRUCTION 10.2. *For a Hurwitz cover* (S, f), *we can associate a marked surface* $\mathbf{S}^f = (\mathrm{S}^f, \mathbf{M}, \mathbf{Y})$ *as follows:*

- *A surface* S^f *is defined as the real blow up of the underlying smooth surface of* S *at the poles* $f^{-1}(\infty) = \{z_1, \dots, z_b\}$.
- *Define marked arcs* $\overline{\mathbf{M}} \subset \partial \mathrm{S}^f$ *as open intervals corresponding to asymptotic directions, where* e^f *decays exponentially as* $z \to z_i$.

- *By shrinking each marked arc of* $\overline{\mathbf{M}}$ *to a single point in* S^f*, we obtain a set of closed marked points* $\mathbf{M}$*.*
- *The set of open marked points in* $\mathbf{Y} \subset \partial \mathrm{S}^f$ *are the dual to* $\mathbf{M}$*, i.e., they are alternative in* $\partial \mathrm{S}^f$*.*

Denote by ∂_i the S^1-boundary component obtained from the real blow up at z_i for $i = 1, \dots, b$. Then by definition, there are $k_i = -\operatorname{ord}_f(z_i)$ points on the boundary component ∂_i.

REMARK 10.3. Though we can construct the marked surface $\mathrm{S}^f = (\mathrm{S}^f, \mathbf{M}, \mathbf{Y})$ from the Hurwitz cover (S, f), this information is not enough to construct a grading λ on S^f. In Section 10.5, we introduce a primary quadratic differential in the case when the genus of S^f is zero, which allows us construct a natural grading.

Next we consider the construction of log surfaces from Hurwitz covers. Let (S, f) be a regular Hurwitz cover. Consider a multi-valued holomorphic function $\log f$ on $\mathrm{S}^\phi := \mathrm{S} \setminus \mathrm{Crit}(f)$. Take a base point $* \in \mathrm{S}^\phi$ and choose a value $\log f(*)$. First we note that the analytic continuation of $\log f(*)$ along with a cycle $\gamma \in \pi_1(\mathrm{S}^\phi, *)$ is given as the form $\log f(*) + 2\pi m$ for some $m \in \mathbb{Z}$. Then we obtain a representation

$$\rho \colon \pi_1(\mathrm{S}^\phi, *) \to \langle q \rangle, \quad \gamma \mapsto \rho(\gamma) := q^m.$$

DEFINITION 10.4. The *logarithmic (*log*) surface* $\log \mathrm{S}^\phi$ associated to a Hurwitz cover (S, f) is defined to be the covering $\pi \colon \log \mathrm{S}^\phi \to \mathrm{S}^\phi$ corresponding to the representation $\rho \colon \pi_1(\mathrm{S}^\phi, *) \to \langle q \rangle$. The infinite cyclic group $\langle q \rangle$ acts on $\log S^\circ$ as the group of deck transformations.

By definition, the multi-valued function $\log f$ can be lifted on the log surface $\log \mathrm{S}^\phi$ as a single-valued function through the analytic continuation. In other words, the surface $\log \mathrm{S}^\phi$ is the Riemann surface of the function $\log f$. We also note that for any complex number $s \in \mathbb{C}$, the function

$$f^{s-2} = e^{(s-2)\log f}$$

is also a single-valued function on $\log \mathrm{S}^\phi$.

Finally we note that as in Section 7.4, for a log surface $\log \mathrm{S}^\phi$ we can construct the topological log surface $\log \mathrm{S}^f_\Delta$ by adding zeroes Δ of f and doing the real blow-ups at poles of f.

10.3. Primary quadratic differentials

In the following, we assume $g = 0$, namely $\mathrm{S} \cong \mathbb{P}^1$, and fix the coordinate $z \in \mathbb{C} \cup \{\infty\} = \mathbb{P}^1$.

DEFINITION 10.5. Let $\mathbf{l} = (l_1, \dots, l_b) \in \mathbb{Z}^b$ be an ordered multi-set of b integers satisfying the condition

$$\sum_{i=1}^{b} l_i = 4.$$

For a Hurwitz cover $(\overline{\mathbb{C}}, f)$ with poles $f^{-1}(\infty) = \{z_1, \dots, z_b\}$, we define a meromorphic quadratic differential $\Omega_\mathbf{l}$ by

$$\Omega_\mathbf{l}(z) := \prod_{i=1}^{b} \frac{\mathrm{d}\, z^2}{(z - z_i)^{l_i}}. \tag{10.1}$$

We call $\Omega_{\mathbf{l}}$ the *primary quadratic differential compatible with* $(\overline{\mathbb{C}}, f)$.

For a Hurwitz cover $f\colon \overline{\mathbb{C}} \to \overline{\mathbb{C}}$ with poles of order $k_1, \dots, k_b$ at $z_1, \dots, z_b$ respectively, consider the primary quadratic differential $\Omega_{\mathbf{l}}$. Then the quadratic differential

$$\phi = \phi(f, \mathbf{l})\colon = e^f \Omega_{\mathbf{l}} \tag{10.2}$$

on $\overline{\mathbb{C}}$ gives a quadratic differential with exponential type singularities of indexes

$$(\mathbf{k}, \mathbf{l}) = \{(k_1, l_1), \dots, (k_b, l_b)\}$$

at z_1, ..., z_b respectively. The following statement gives the other direction.

PROPOSITION 10.6. *Let ϕ be a quadratic differential on $\overline{\mathbb{C}}$ with exponential type singularities of indexes $(\mathbf{k}, \mathbf{l})$ at $z_1, \dots, z_b$ respectively. Then there is a unique Hurwitz cover $f\colon \overline{\mathbb{C}} \to \overline{\mathbb{C}}$ with poles of order $k_1, \dots, k_b$ at $z_1, \dots, z_b$, up to $f \mapsto f + 2\pi \mathbf{i} m$ for $m \in \mathbb{Z}$, such that ϕ is of the form in* (10.2).

PROOF. First we consider the holomorphic function $g(z) := \phi(z)/\Omega_{\mathbf{l}}(z)$ on $\overline{\mathbb{C}} \setminus \{z_1, \dots, z_b\}$. Then by definition, $g(z)$ is non-zero on $\overline{\mathbb{C}} \setminus \{z_1, \dots, z_b\}$ and points $z_1, \dots, z_b$ are exponential type singularities. Since $g(z)$ is non-zero away from z_i, it is sufficient to show that the function $\log g(z)$ has a pole of order k_i at z_i and is single-valued around z_i. By definition of exponential type singularity, there is a coordinate change $v(z)$ satisfying $v(z_i) = 0$ and $v'(z_i) \neq 0$ near z_i such that

$$\phi = e^{v^{-k_i}} v^{-l_i} \cdot h_1(v) \, \mathrm{d} v^2,$$

where $h_1(v)$ is some non-zero holomorphic function near $v = 0$. We also note that under the above coordinate change, the primary quadratic differential $\Omega_{\mathbf{l}}$ can be written as

$$\Omega_{\mathbf{l}} = v^{-l_i} \cdot h_2(v) \, \mathrm{d} v^2,$$

where $h_2(v)$ is also some non-zero holomorphic function near $v = 0$. Thus $g(z(v)) = \exp(v^{-k_i}) \cdot h_1(v)/h_2(v)$ near $v = 0$. Thus $\log g$ has a pole of order k_i at z_i as required. □

Note that the primary quadratic differential $\phi = \phi(f, \mathbf{l})$ in (10.2) provides a canonical grading $\lambda = \lambda(f, \mathbf{l})$ on the associated marked surface S^f. Define the numerical data for the Hurwitz cover $(\overline{\mathbb{C}}, f)$ by

$$\mathfrak{N}(\overline{\mathbb{C}}, f) = \mathfrak{N}\left((\mathrm{S}^f)^\lambda\right). \tag{10.3}$$

10.4. q-Quadratic differentials

Similar to Proposition 10.6, we consider the analogue statement for CY-s type quadratic differentials. Fix a complex number s with $\mathrm{Re}(s) > 2$. For a Hurwitz cover $f\colon \overline{\mathbb{C}} \to \overline{\mathbb{C}}$ with simple zeroes and poles of order $k_1, \dots, k_b$ at $z_1, \dots, z_b$ respectively, define a multi-valued quadratic differential on $\overline{\mathbb{C}}$ ramified at $\mathrm{Crit}(f)$ by

$$\xi = \xi(f, s, \mathbf{l})\colon = f^{s-2} \Omega_{\mathbf{l}} = e^{(s-2) \log f} \Omega_{\mathbf{l}} \tag{10.4}$$

Then ξ is a CY-s type q_s-quadratic differential with s-poles of type $(\mathbf{k}, \mathbf{l})$ at $z_1, \dots$, z_b. The following statement gives the other direction in this case.

PROPOSITION 10.7. *Let ξ be a CY-s type q_s-quadratic differential on $\overline{\mathbb{C}}$ with s-poles of type $(k_1, l_1), \ldots, (k_b, l_b)$ at $z_1, \ldots, z_b$ together with a fixed value $\xi(*)$ on some point $* \in \log(\overline{\mathbb{C}})^\circ$. Then there is a unique regular Hurwitz cover $f\colon \overline{\mathbb{C}} \to \overline{\mathbb{C}}$ with poles of order $k_1, \ldots, k_b$ at $z_1, \ldots, z_b$ and a single-valued function $\log f$ on $\log \overline{\mathbb{C}}^\circ$, up to multiplying by ω_{s-2}^m, for $\omega_{s-2} = e^{2\pi \mathbf{i}/(s-2)}$ and $m \in \mathbb{Z}$:*

$$f \mapsto \omega_{s-2}^m f \quad \text{or} \quad \log f \mapsto \log f + \frac{2\pi \mathbf{i} m}{s-2}, \tag{10.5}$$

such that

$$\xi = f^{s-2}\,\Omega_{\mathbf{l}},$$

where $\mathbf{l} = (l_1, \ldots, l_b)$ *and the value* $\log f(*)$ *is determined by* $\xi(*) = e^{(s-2)\log f(*)}\Omega_{\mathbf{l}}(*)$.

PROOF. First we note that the ambiguity $f \mapsto \omega_{s-2}^m f$ comes from $f^{s-2} = (\omega_{s-2}^m f)^{s-2}$ since $\omega_{s-2}^{s-2} = 1$. Consider a multi-valued function $g(z) := \xi(z)/\Omega_{\mathbf{l}}(z)$ and let

$$f(z) \colon = g(z)^{1/(s-2)} = e^{\frac{1}{s-2}\log g(z)}$$

up to multiplying ω_{s-2}^m. We only need to show that $f(z)$ has a pole of order k_i at z_i and is single-valued near z_i.

Similar to the proof of Proposition 10.6, there is a coordinate change $v(z)$ satisfying $v(z_i) = 0$ and $v'(z_i) \neq 0$ near z_i such that

$$\xi = v^{-k_i(s-2)-l_i}\,\mathrm{d}\,v^2$$

and

$$\Omega_{\mathbf{l}} = v^{-l_i} \cdot h(v)\,\mathrm{d}\,v^2,$$

where $h(v)$ is also some non-zero holomorphic function near $v = 0$. Then $g(z(v)) = v^{-k_i(s-2)} \cdot h(v)^{-1}$ near $v = 0$. Thus $f(z(v)) = g(z(v))^{1/(s-2)}$ has a pole of order k_i at $v = 0$ and is single-valued near $v = 0$ since $h(v)^{-1}$ is non-zero. □

10.5. Framed regular Hurwitz spaces

Let $\mathbf{C}^\lambda = (\mathbf{C}, \mathbf{M}, \mathbf{Y}, \lambda)$ be a graded marked surface of genus zero. Then the numerical data of $\mathbf{C}$ is just

$$\mathfrak{N}(\mathbf{C}^\lambda) = (b; \mathbf{k}, \mathbf{l})$$

since $g = 0$ and LP_0 is empty. We also choose an arc system $\mathbf{A}$ and a cut $\mathbf{c}$ and thus have the log surface $\log_{\mathbf{c}} \mathbf{C}_\Delta^\lambda$, which will be simply written as $\log \mathbf{C}_\Delta$.

Let $(\overline{\mathbb{C}}, f)$ be a regular Hurwitz cover of polar type $\mathbf{k}$ and $\log \overline{\mathbb{C}}_\Delta^f$ be the associated topological log surface constructed in Section 10.2 with numerical data $\mathfrak{N}(\overline{\mathbb{C}}, f)$ in (10.3). Then Theorem 7.21 implies that there is a homeomorphism

$$h\colon \log \mathbf{C}_\Delta \to \log \overline{\mathbb{C}}_\Delta^f$$

which commutes with the action of $\langle q \rangle$.

DEFINITION 10.8. A $\log \mathbf{C}_\Delta$*-framed Hurwitz cover* $(\overline{\mathbb{C}}, f, \log f, h)$ consists of a regular Hurwitz cover $(\overline{\mathbb{C}}, f)$, a fixed choice of the (single-valued) function $\log f$ on $\log \overline{\mathbb{C}}^\circ$ and a homeomorphism

$$h\colon \log \mathbf{C}_\Delta \to (\log \overline{\mathbb{C}}_\Delta^\circ)^f,$$

which commutes with the action of $\langle q \rangle$ and matches the numerical data $\mathfrak{N}(\overline{\mathbb{C}}, f)$ with $\mathfrak{N}(\log \mathbf{C}_\Delta)$.

Two $\log \mathbf{C}_\Delta$-framed Hurwitz cover $(\overline{\mathbb{C}}, f_i, \log f_i, h_i)$ are $\log \mathbf{C}_\Delta$-equivalent if

- there is a biholomorphism $F\colon \overline{\mathbb{C}} \to \overline{\mathbb{C}}$ satisfying $F(f_1^{-1}(\infty)) = f_2^{-1}(\infty)$, together with the choice of a lift $\widetilde{F}\colon \log \overline{\mathbb{C}}^\circ \to \log \overline{\mathbb{C}}^\circ$ such that $\widetilde{F}^*(\log f_2) = \log f_1$ and
- $h_2^{-1} \circ \widetilde{F} \circ h_1 \in \mathrm{Homeo}_0(\log \mathbf{C}_\Delta)$, where

$$\widetilde{F}\colon \log \overline{\mathbb{C}}_\Delta^{f_1} \to \log \overline{\mathbb{C}}_\Delta^{f_2}$$

is the induced homeomorphism.

Denote by $\mathrm{HS}(\log \mathbf{C}_\Delta)$ the moduli space of $\log \mathbf{C}_\Delta$-framed Hurwitz covers.

Recall that $\mathrm{MCG}^\circ(\log \mathbf{C}_\Delta)$ is generated by the deck transformation q and those homeomorphism induced from $\mathrm{MCG}(\mathbf{C}_\Delta)$. The deck transformation $q \in \mathrm{MCG}^\circ(\log \mathbf{C}_\Delta)$ acts on $\mathrm{HS}(\log \mathbf{C}_\Delta)$ as

$$(\overline{\mathbb{C}}, f, \log f, h) \cdot q = (\overline{\mathbb{C}}, f, \log f + 2\pi \mathbf{i}, h).$$

The right action of $g \in \mathrm{MCG}^\circ(\log \mathbf{C}_\Delta)$ on $\mathrm{HS}(\log \mathbf{C}_\Delta)$ is given by

$$(\overline{\mathbb{C}}, f, \log f, h) \cdot g = (\overline{\mathbb{C}}, f, \log f, h \circ g).$$

Choose a boundary component $\partial_1 \subset \mathbf{C}_\Delta$ with k_1 marked points and set $\widetilde{\partial}_1 := \pi^{-1}(\partial_1)$ for the covering map $\pi_\Delta\colon \log \mathbf{C}_\Delta \to \mathbf{C}_\Delta$. For a $\log \mathbf{C}_\Delta$-framed Hurwitz cover $(\overline{\mathbb{C}}, f, \log f, h)$, assume that a pole z_1 of f corresponds to $\widetilde{\partial}_1$ through the framing h. Then by the automorphism of $\overline{\mathbb{C}}$, we can always take the representative $(\overline{\mathbb{C}}, f, \log f, h)$ with $z_1 = \infty$ and the coefficient of the pole of f at $z_1 = \infty$ is one, namely

$$f = z^{k_1} + O(z^{k_1+1}). \tag{10.6}$$

Recall from Definition 7.28 that $\mathrm{QQuad}_s(\log \mathbf{C}_\Delta)$ is the moduli space of q-quadratic differentials with the numerical data $\mathfrak{N}(\log \mathbf{C}_\Delta) = (b, \mathbf{k}, \mathbf{l})$.

THEOREM 10.9. *There is an isomorphism of complex manifolds*

$$\begin{array}{rccc} \Psi_s\colon & \mathrm{HS}(\log \mathbf{C}_\Delta) & \longrightarrow & \mathrm{QQuad}_s(\log \mathbf{C}_\Delta) \\ & (\overline{\mathbb{C}}, f, \log f, h) & \longmapsto & (\overline{\mathbb{C}}, \xi, h, s) \end{array} \tag{10.7}$$

for $\xi = \xi(f, x, \mathbf{l})$ *as in* (10.4). *Here we use the representative* f *with the form* (10.6).

PROOF. The existence and injectivity of the map Ψ_s are explicit. We only need to show the surjectivity of Ψ_s. For a q-quadratic differential, by the automorphism of $\overline{\mathbb{C}}$, we can take its representative as $(\overline{\mathbb{C}}, \xi, h, s)$ with $z_1 = \infty$, where z_1 is an s-pole corresponding to $\widetilde{\partial}_1$. In addition, we can normalize as

$$\xi = c(z^{k_1(s-2)} + \cdots)\Omega_{\mathbf{l}},$$

where $c = q_s^{2m}$ for some $m \in \mathbb{Z}$. Take some point $*$ from the interior of $\log \mathbf{C}_\Delta$. Then by Proposition 10.7, there is a regular Hurwitz cover $f\colon \overline{\mathbb{C}} \to \overline{\mathbb{C}}$ and a single-valued function $\log f$ on $\log \overline{\mathbb{C}}^\circ$, up to multiplying by ω_{s-2}^m as in (10.5), such that

$$\xi = f^{s-2}\Omega_{\mathbf{l}} \quad \text{and} \quad \xi(h(*)) = e^{(s-2)\log f(h(*))}\Omega_{\mathbf{l}}(h(*)).$$

By the construction of f in the proof of Proposition 10.7, we can take f with the form $f = z^{k_1} + \cdots$. This implies the surjectivity of Ψ_s. $\square$

REMARK 10.10. As a complex manifold, the moduli space $\mathrm{QQuad}_s(\log \mathbf{C}_\Delta)$ is independent of the date $\mathbf{l}$ of $\mathfrak{N}(\log \mathbf{C}_\Delta)$. However, the period map

$$\Pi_s = \int \sqrt{\xi}$$

strongly depends on $\mathbf{l}$ since $\xi = f^{s-2}\Omega_{\mathbf{l}}$. The choice $\mathbf{l}$ corresponds to the choice of the primitive form [**Sa1**, **SaTa**] (the primary differential [**D1**, Section 5]) and controls the Frobenius structure on $\mathrm{HS}(\log \mathbf{C}_\Delta)$.

COROLLARY 10.11. *Consider the union*

$$\mathrm{QQuad}_{>2}(\log \mathbf{C}_\Delta) \colon = \bigcup_{s \in \mathbb{C}^{>2}} \mathrm{QQuad}_s(\log \mathbf{C}_\Delta),$$

where $\mathbb{C}^{>2} = \{s \in \mathbb{C} \mid \mathrm{Re}(s) > 2\}$. *Then the space* $\mathrm{QQuad}(\log \mathbf{C}_\Delta)$ *has the structure of a complex manifold of dimension* $(n+1)$ *induced from the bijection*

$$\mathrm{QQuad}_{>2}(\log \mathbf{C}_\Delta) \cong \mathrm{HS}(\log \mathbf{C}_\Delta) \times \mathbb{C}^{>2}$$

given by Theorem 10.9.

Passing to q-stability conditions, we have the following result on gluing q-stability conditions along s-direction for

$$\mathrm{QStab}^\circ_{>2} \mathcal{D}_{\mathbb{X}}(\mathbf{A}) = \bigcup_{s \in \mathbb{C}^{>2}} \mathrm{QStab}^\circ_s \mathcal{D}_{\mathbb{X}}(\mathbf{A}), \tag{10.8}$$

which provides an answer for [**IQ**, Conj. 3.16].

COROLLARY 10.12. *For any s with* $\mathrm{Re}(s) > 2$, *we have*

$$\Xi_s \circ \Psi_s \colon \mathrm{HS}(\log \mathbf{C}_\Delta) \cong \mathrm{QStab}^\circ_s \mathcal{D}_{\mathbb{X}}(\mathbf{A}), \tag{10.9}$$

where Ξ_s *and* Ψ_s *are as in* (8.9) *and* (10.7). *Hence* $\mathrm{QStab}^\circ_{>2} \mathcal{D}_{\mathbb{X}}$ *admits the structure of a complex manifold of dimension* $n+1$ *and the projection map*

$$\pi \colon \mathrm{QStab}^\circ \mathcal{D}_{\mathbb{X}} \to \mathbb{C}, \quad (\sigma, s) \mapsto s$$

is holomorphic.

PROOF. In our case, $\mathbf{S} = \overline{\mathbb{C}}$ has genus zero and thus Corollary 8.7 applies. By Theorem 8.9, we have (8.12), i.e., $\mathrm{QStab}^\circ_s \mathcal{D}_{\mathbb{X}}(\mathbf{A}) \cong \mathrm{QQuad}_s(\log \mathbf{C}_\Delta)$ in this case. Thus the claim follows from Theorem 8.4 and Corollary 10.11 respectively. □

10.6. Disk case: Type A example

Now consider the type A case, where $\mathrm{S} = \overline{\mathbb{C}}$, $b = 1$, $\mathbf{k} = (n+1)$ and $\mathbf{l} = (4)$. In this case, the marked surface $\mathbf{C}$ is a disk with $n+1$ points on the boundary, which will be denoted by $\mathbf{D}$. Consider the space of polynomials of degree $n+1$

$$\mathcal{M}_n := \{f(z) \mid f(z) = z^{n+1} + a_1 z^{n-1} + a_2 z^{n-2} + \cdots + a_n,\ a_i \in \mathbb{C}\} \cong \mathbb{C}^n \tag{10.10}$$

and its regular subset

$$\mathcal{M}_{n,\mathrm{reg}} := \{f(z) \in \mathcal{M}_n \mid \text{all zeroes of } f(z) \text{ are simple}\}.$$

Then we can regard $f(z) \in \mathcal{M}_{n,\mathrm{reg}}$ as a genus 0 regular Hurwitz cover

$$f \colon \overline{\mathbb{C}} \to \overline{\mathbb{C}}$$

with a unique pole of order $n+1$ at $z = \infty$.

Let $\mathfrak{h}$ be the Cartan subalgebra of type A_n and $\mathfrak{h}_{\mathrm{reg}} \subset \mathfrak{h}$ be the regular subset. Then there is an identification

$$\mathcal{M}_{n,\mathrm{reg}} \cong \mathfrak{h}_{\mathrm{reg}}/W,$$

where W is the Weyl group of type A_n, namely the symmetric group of degree $n+1$. More details on $\mathfrak{h}$ in the Dynkin case will be discussed in Section 11.3. Denote by $\widetilde{\mathcal{M}}_{n,\mathrm{reg}}$ the universal cover of $\mathcal{M}_{n,\mathrm{reg}}$.

Let $\omega_{n+1} = e^{2\pi \mathbf{i}/(n+1)}$ be the $(n+1)$-th root of unity and denote by $C_{n+1} := \langle \omega_{n+1} \rangle$ the cyclic group of order $n+1$ generated by ω_{n+1}. We define the action of C_{n+1} on $\mathcal{M}_{n,\mathrm{reg}}$ by

$$(\omega_{n+1} \cdot f)(z) := f(\omega_{n+1}^{-1} z).$$

Alternatively, we can write $\omega_{n+1} \cdot (a_i)_i = (\omega_{n+1}^{i+1} a_i)_i$ for coefficients $(a_i)_i$ of $f(z)$.

LEMMA 10.13. *There is an isomorphism*

$$\mathcal{M}_{n,\mathrm{reg}}/C_{n+1} \cong \mathrm{HS}(0,(n+1))_{\mathrm{reg}} \tag{10.11}$$

by regarding a regular polynomial as a regular Hurwitz cover on $\overline{\mathbb{C}}$.

PROOF. It is straightforward to check that the only automorphisms of $\overline{\mathbb{C}}$ which preserve $\mathcal{M}_{n,\mathrm{reg}}$ are $z \mapsto \omega z$ for $\omega \in C_{n+1}$. Hence the claim follows. □

We note that the cyclic group C_{n+1} can be identified with the *marked mapping class group* $\mathrm{MCG}_\bullet(\mathbf{D})$, whose generator is the rotation τ that sends each marked point on the boundary to the next (clockwise) point. Recall that $\mathrm{MCG}_\bullet$ here is only required to preserve $\mathbf{M}$ setwise. Let $\mathbf{D}_\Delta$ be the decorated version of $\mathbf{D}$ (with $|\Delta| = n+1$). It is well-known that

$$\mathrm{MCG}_\bullet(\mathbf{D}_\Delta) = (\mathrm{Br}_{n+1} \times \langle \tau \rangle) / \langle \tau^{n+1} \rangle,$$

which follows from the short exact sequence

$$0 \to \mathrm{SBr}(\mathbf{D}_\Delta) \to \mathrm{MCG}_\bullet(\mathbf{D}_\Delta) \to \mathrm{MCG}_\bullet(\mathbf{D}) \to 0$$

for

$$\mathrm{SBr}(\mathbf{D}_\Delta) = \mathrm{MCG}(\mathbf{D}_\Delta, \partial \mathbf{D}_\Delta) = \mathrm{Br}_{n+1}.$$

Here, $\mathrm{MCG}(\mathbf{D}_\Delta, \partial \mathbf{D}_\Delta)$ is the classical mapping class group that preserves the boundary pointwise. More precisely, the τ^{n+1} is the Dehn twist along $\partial \mathbf{D}_\Delta$, which is also the generator of the center of Br_{n+1}.

Let $\log \mathbf{D}_\Delta$ be the topological (ungraded) log surface of type A. Then first we note that the lift of the action $\tau \in \mathrm{MCG}_\bullet(\mathbf{D})$ on $\log \mathbf{D}_\Delta$ can be identified with the deck transformation $q \in \mathrm{MCG}^\circ(\log \mathbf{D}_\Delta)$. Then we have the following description of the mapping group $\mathrm{MCG}^\circ(\log \mathbf{D}_\Delta)$.

LEMMA 10.14. *As*

$$\mathrm{MCG}^\circ(\log \mathbf{D}_\Delta) = (\mathrm{Br}_{n+1} \times \langle q \rangle) / \langle q^{n+1} \rangle,$$

where q^{n+1} *acts on* Br_{n+1} *as the lift of* $D_{\partial \mathbf{D}_\Delta}$, *there is an isomorphism of the groups*

$$\mathrm{MCG}^\circ(\log \mathbf{D}_\Delta) \cong \mathrm{Br}_{n+1} \rtimes C_{n+1}.$$

PROPOSITION 10.15. *There is an isomorphism of complex manifolds*

$$\widetilde{\Phi}_{n+1} \colon \widetilde{\mathcal{M}}_{n,\mathrm{reg}} \to \mathrm{HS}(\log \mathbf{D}_\Delta), \tag{10.12}$$

as universal covering of

$$\begin{array}{rcl} \Phi_{n+1}\colon \mathrm{HS}(0,(n+1))_{\mathrm{reg}} & \longrightarrow & \mathrm{HS}(\log \mathbf{D}_\Delta)/\operatorname{MCG}^\circ(\log \mathbf{D}_\Delta) \\ (\overline{\mathbb{C}}, f) & \longmapsto & (\overline{\mathbb{C}}, f, \log f, h). \end{array} \tag{10.13}$$

PROOF. By Definition, we have (10.13).

Let

$$\mathrm{conf}_m(X)\colon = \{\{x_1, \ldots, x_n\} \in X^n \mid x_i \neq x_j,\ \forall i \neq j\}$$

be the mth (unordered) configuration space of X. Then there is a natural identification

$$\begin{aligned} \mathcal{M}_{n,\mathrm{reg}} &= \{f(z) = \prod_{k=1}^{n+1}(z - z_k) \mid z_k \in \mathbb{C}, z_i \neq z_j,\ \forall i \neq j\} \\ &= \mathrm{conf}_{n+1}(\mathbb{C}). \end{aligned}$$

It is well-known that $\pi_1(\mathcal{M}_{n,\mathrm{reg}}) = \mathrm{Br}_{n+1}$. By Lemma 10.13, we have

$$\pi_1(\mathrm{HS}(0,(n+1))_{\mathrm{reg}}) = \pi_1(\mathcal{M}_{n,\mathrm{reg}}/C_{n+1}) = \mathrm{Br}_{n+1} \rtimes C_{n+1}.$$

Combining Lemma 10.14, we obtain the required covering version (10.12). $\square$

CHAPTER 11

Almost Frobenius structures for stability conditions: ADE case

The aim of this section is to give a conjectural description of the spaces of q-stability conditions on the Calabi-Yau-$\mathbb{X}$ completions of type ADE quivers.

11.1. Almost Frobenius structures

We recall the definition of an almost Frobenius structure from [**D1**, § 3].

DEFINITION 11.1. A *Frobenius algebra* $(A, *, (\ ,\ ))$ is a pair of a commutative associative $\mathbb{C}$-algebra $(A, *)$ with the unit $E \in A$ and a non-degenerate symmetric bilinear form

$$(\ ,\ ) \colon A \otimes A \to \mathbb{C}$$

satisfying the condition

$$(X * Y, Z) = (X, Y * Z)$$

for $X, Y, Z \in A$.

For a complex manifold M, denote by $\mathcal{T}_M$ the holomorphic tangent sheaf of M.

DEFINITION 11.2. An *almost Frobenius manifold* $(M, *, (\ ,\ ), E, e)$ of the charge $d \in \mathbb{C}\,(d \neq 1)$ consists of the following data:

- a complex manifold M of dimension n,
- a commutative associative $\mathcal{O}_M$-linear product

$$* \colon \mathcal{T}_M \otimes \mathcal{T}_M \to \mathcal{T}_M$$

- a non-degenerate symmetric $\mathcal{O}_M$-bilinear form

$$(\ ,\ ) \colon \mathcal{T}_M \otimes \mathcal{T}_M \to \mathcal{O}_M$$

  satisfying the condition

$$(X * Y, Z) = (X, Y * Z)$$

  for (local) holomorphic vector fields X, Y, Z,
- global holomorphic vector fields E (called the *Euler vector field*) and e (called the *primitive vector field*).

They satisfy the following axioms:

(1) The metric $(\ ,\ )$ is flat.

(2) Let $p^1, \dots, p^n$ be flat coordinates of the flat metric $(\ ,\ )$ on some open subset $U \subset M$ and set $\partial_i := \frac{\partial}{\partial p^i}$. Then there is some function $F(p) \in \mathcal{O}_M(U)$ such that the multiplication $*$ can be written as

$$\partial_i * \partial_j = \sum_{k=1}^{n} C_{ij}^k(p)\partial_k,$$

where

$$C_{ij}^k(p) = \sum_{l=1}^{n} G^{kl}\partial_i\partial_j\partial_l F(p)$$

and (G^{kl}) is the inverse matrix of the metric $(G_{kl} = (\partial_k, \partial_l))$.

(3) The function $F(p)$ satisfies the homogeneity equation

$$\sum_{i=1}^{n} p^i\partial_i F(p) = 2F(p) + \frac{1}{1-d}\sum_{i,j=1}^{n} G_{ij}p^ip^j.$$

(4) The Euler vector field takes the form

$$E = \frac{1-d}{2}\sum_{i=1}^{n} p^i\partial_i$$

and is the unit of the multiplication $*$.

(5) The primitive vector field e is invertible with respect to the multiplication $*$ and the derivation $P_\nu(p) \mapsto eP_\nu(p)$ on the space of solutions of the following integrable equation with the parameter $\nu \in \mathbb{C}$

$$\partial_i\partial_j P_\nu(p) = \nu\sum_{k=1}^{n} C_{ij}^k(p)\partial_k P_\nu(p) \tag{11.1}$$

acts as the shift of the parameter $\nu \mapsto \nu - 1$, i.e., if $P_\nu(p)$ is the solution of the equation (11.1) with the parameter ν, then $eP_\nu(p)$ is the solution with the parameter $\nu - 1$.

We note that the restriction of the multiplication and the metric on the tangent space T_pM for each point $p \in M$ gives a family of Frobenius algebras $(T_pM, *_p, (\ ,\ )_p)$.

The flat coordinates $p^1, \dots, p^n$ of the flat metric $(\ ,\ )$ are called *periods* and independent solutions $P_\nu^1, \dots, P_\nu^n$ of the equation in Axiom (5) are called *twisted periods*. The twisted periods satisfy the following quasi-homogeneity condition.

PROPOSITION 11.3. *The derivative of the twisted period P_ν along with the Euler field E satisfies*

$$EP_\nu = \left(\frac{1-d}{2} + \nu\right)P_\nu + \text{const}.$$

PROOF. We show

$$\partial_j EP_\nu = \left(\frac{1-d}{2} + \nu\right)\partial_j P_\nu.$$

Since $[\partial_j, E] = \frac{1-d}{2}\partial_j$, the above equality is equivalent to $E\partial_j P_\nu = \nu\partial_j P_\nu$. Recall that the Euler field is the unit of the Frobenius algebra, i.e. $E * \partial_j = \partial_j$. This is equivalent to

$$\frac{1-d}{2}\sum_{i,k=1}^{n} p^i C_{ij}^k\partial_k = \partial_j.$$

Together with the equation (11.1) for the twisted period $\partial_i \partial_j P_\nu(p) = \nu \sum_k C_{ij}^k \partial_k P_\nu$, we have

$$E\partial_j P_\nu = \frac{1-d}{2}\sum_{i=1}^{n} p^i \partial_i \partial_j P_\nu = \nu \frac{1-d}{2}\sum_{i,k=1}^{n} p^i C_{ij}^k \partial_k P_\nu = \nu \partial_j P_\nu.$$

□

In the following, we always normalize the twisted period P_ν to satisfy the quasi-homogeneity condition

$$EP_\nu = \left(\frac{1-d}{2} + \nu\right) P_\nu \tag{11.2}$$

by shifting some constant.

11.2. Flat torsion-free connections and twisted periods

In this section, we summarize the relation between flat torsion-free connections and affine structures on manifolds. In the following we consider in the category of complex manifolds and all vector fields, differential forms and connections are holomorphic. Let M be a complex manifold of dimension n and ∇ be a connection of the holomorphic tangent bundle TM. We recall that ∇ is called *torsion-free* if

$$\nabla_X Y - \nabla_Y X - [X, Y] = 0$$

for vector fields X, Y on M. Recall that a connection ∇ on the tangent bundle TM induces the connection ∇ on the cotangent bundle T^*M by the formula

$$\mathrm{d}\langle X, \omega\rangle = \langle \nabla X, \omega\rangle + \langle X, \nabla\omega\rangle$$

for a vector field X and a 1-form ω on M. The following statement is well-known.

PROPOSITION 11.4. *Let M be a simply connected complex manifold of dimension n and ∇ be a flat torsion-free connection on TM. Then there are n holomorphic functions $P^1, \dots, P^n \colon M \to \mathbb{C}$ such that*

$$\mathrm{d}\,P^1 \wedge \cdots \wedge \mathrm{d}\,P^n \neq 0$$

everywhere on M and

$$\nabla\,\mathrm{d}\,P^i = 0.$$

These functions are unique up to affine transformation

$$P^i \mapsto \sum_{j=1}^{n} A^i_j P^j + B^i,$$

where $(A^i_j) \in \mathrm{GL}(n, \mathbb{C})$ and $(B^i) \in \mathbb{C}^n$.

Thus functions $P^1, \dots, P^n$ give local coordinates on M. This coordinate system is called a *flat coordinate system.* Consider the vector space

$$V_M := \bigoplus_{i=1}^{n} \mathbb{C}\,\mathrm{d}\,P^i.$$

Then since $\mathrm{d}\,P^1, \dots, \mathrm{d}\,P^n$ are independent global sections of T^*M, we have a trivialization of the cotangent bundle

$$T^*M \cong V_M \times M$$

and the flat torsion-free connection ∇ is just equal to the exterior differential d on the trivial bundle.

Consider an almost Frobenius manifold $(M, *, (\ ,\ ), E, e)$. Let ∇ be the Levi-Civita connection with respect to the flat metric $(\ ,\ )$ on M. Fix a complex number $\nu \in \mathbb{C}$. Then we can define a flat torsion-free connection $\check{\nabla}$ on TM

$$\check{\nabla}_X Y := \nabla_X Y + \nu X * Y$$

for vector fields X, Y. Then the equation (11.1) for twisted periods P_ν is equivalent to the equation

$$\check{\nabla}\, \mathrm{d}\, P_\nu = 0.$$

REMARK 11.5. The connection $\check{\nabla}$ is called the *Dubrovin connection* or the *second structure connection* of a Frobenius manifold. For details, we refer to [**D1**, § 3] and [**M**, § 2.1].

11.3. ADE type

We recall the almost Frobenius structure on the universal covering of the regular locus of the Cartan subalgebra in the case of type ADE following [**D1**, § 5.2].

Let Q be a quiver of type ADE and $(L_Q, (\ ,\ ))$ be the root lattice associated with Q. Let $\Delta_+ \subset L_Q$ be the set of positive roots. The Cartan subalgebra is

$$\mathfrak{h} := \mathrm{Hom}_{\mathbb{Z}}(L_Q, \mathbb{C})$$

and let $e_1, \dots, e_n \in \mathfrak{h}$ be the dual basis of $\alpha_1, \dots, \alpha_n$. Then

$$\mathfrak{h} \cong \oplus_{i=1}^n \mathbb{C} e_i \quad \text{with metric} \quad (e_i, e_j) := (A^{-1})_{ij},$$

where A^{-1} is the inverse of the Cartan matrix. We regard a root $\alpha \in \Delta_+$ as a function on $\mathfrak{h}$ via the pairing between $\mathfrak{h}$ and L_Q. Consider the regular subset

$$\mathfrak{h}_{\mathrm{reg}} := \mathfrak{h} \setminus \bigcup_{\alpha \in \Delta_+} \ker \alpha \tag{11.3}$$

and its universal covering space

$$\widetilde{\mathcal{M}}_Q := \widetilde{\mathfrak{h}_{\mathrm{reg}}}.$$

When Q is an A_n quiver, these definitions/notations match the ones in Section 10.6. In the following, we construct the almost Frobenius structure on $\widetilde{\mathcal{M}}_Q$.

Let $\alpha_1, \dots, \alpha_n$ be simple roots and introduce the coordinate on $\mathfrak{h}$ given by $p^i := \alpha_i(p)$ for $p \in \mathfrak{h}$, namely $p = \sum_{i=1}^n p^i e_i$. Then the coordinates $p^1, \dots, p^n$ also form the local coordinates on $\widetilde{\mathcal{M}}_Q$ through the projection $\widetilde{\mathcal{M}}_Q \to \mathfrak{h}_{\mathrm{reg}}$. Since $T\widetilde{\mathcal{M}}_Q \cong \mathfrak{h} \times \widetilde{\mathcal{M}}_Q$, the metric $(\ ,\ )$ on $\mathfrak{h}$ induces the flat metric on $\widetilde{\mathcal{M}}_Q$ and the coordinates $p^1, \dots, p^n$ form the flat coordinates.

Define the potential function $F \colon \widetilde{\mathcal{M}}_Q \to \mathbb{C}$ by

$$F(p) := \frac{1}{2} \sum_{\alpha \in \Delta_+} \alpha(p)^2 \log \alpha(p).$$

Since $\widetilde{\mathcal{M}}_Q$ is the universal covering of $\mathfrak{h}_{\mathrm{reg}}$, F is a single-valued holomorphic function on $\widetilde{\mathcal{M}}_Q$. On an open subset $U \subset \widetilde{\mathcal{M}}_Q$ with the flat coordinate $p^1, \dots, p^n$, the third derivative of F is given by

$$C_{ijk}(p) := \partial_i \partial_j \partial_j F(p) = \sum_{\alpha \in \Delta_+} \frac{\alpha(e_i)\alpha(e_j)\alpha(e_k)}{\alpha(p)}. \tag{11.4}$$

This defines the multiplication $*$ together with the inverse of the metric, which is just the Cartan matrix $(dp^i, dp^j) = A_{ij}$. The charge d is given by

$$d = 1 - \frac{2}{h},$$

where h is the Coxeter number. In particular, the Euler field takes the form

$$E = \frac{1}{h} \sum_{i=1}^{n} p^i \partial_i.$$

For the construction of the primitive vector field e, we refer to [**D1**, § 5.2].

11.4. Coxeter-KZ connections

Following [**C**] (and the survey [**TL**, § 4]), we consider a certain integrable connection on $\mathfrak{h}_{\mathrm{reg}}$, called the Coxeter-KZ connection.

As in the previous section, let $(L_Q, (\ ,\ ))$ be the root lattice associated with a quiver Q of type ADE. For a root $\alpha \in \Delta$, consider reflections

$$r_\alpha \colon L_Q \to L_Q, \qquad \lambda \mapsto \lambda - (\lambda, \alpha)\alpha,$$

for $\alpha \in L_Q$. These reflections form the *Weyl group*

$$W_Q := \langle r_\alpha \,|\, \alpha \in \Delta \rangle,$$

which naturally acts on the Cartan subalgebra $\mathfrak{h}$.

Consider a representation of W_Q on a finite dimensional vector space V:

$$\rho \colon W_Q \to \mathrm{GL}(V).$$

On the trivial bundle $V \times \mathfrak{h}_{\mathrm{reg}} \to \mathfrak{h}_{\mathrm{reg}}$, we define the *Coxeter-KZ (CKZ)* connection $\nabla_\nu^{\mathrm{CKZ}}$ by

$$\nabla_\nu^{\mathrm{CKZ}} := \mathrm{d} - \nu \sum_{\alpha \in \Delta_+} (\rho(r_\alpha) - \mathrm{id}_V) \frac{d\alpha}{\alpha},$$

where $\nu \in \mathbb{C}$ is some complex number.

THEOREM 11.6. *The connection $\nabla_\nu^{\mathrm{CKZ}}$ is a W-invariant flat connection on $V \times \mathfrak{h}_{\mathrm{reg}}$. The monodromy representation*

$$\rho_\nu \colon \pi_1(\mathfrak{h}_{\mathrm{reg}}/W) \cong \mathrm{Br}_Q \to \mathrm{GL}(V)$$

factors through the Hecke algebra H_Q:

$$\begin{array}{ccc} R[\mathrm{Br}_Q] & \longmapsto & R[\mathrm{GL}(V)] \\ & \searrow \qquad \nearrow & \\ & H_Q & \end{array} \qquad\qquad \begin{array}{ccc} \psi_i & \longmapsto & \rho_\nu(\psi_i) \\ & \searrow \qquad \nearrow & \\ & -T_i & \end{array} .$$

where we specialize $q = e^{2i\pi\nu}$ for the Hecke algebra H_Q.

PROOF. See [**C**] and also the survey [**TL**, § 4]. □

REMARK 11.7. In [**TL**], the CKZ connection is given by

$$\nabla'^{\mathrm{CKZ}}_{\nu} = \mathrm{d} - \nu \sum_{\alpha\in\Delta_+} \rho(r_\alpha)\frac{d\alpha}{\alpha}.$$

Two connections $\nabla^{\mathrm{CKZ}}_{\nu}$ and $\nabla'^{\mathrm{CKZ}}_{\nu}$ are related as the following gauge transformation. Consider the function

$$f\colon = \prod_{\alpha\in\Delta_+} \alpha$$

on $\mathfrak{h}$. Then the equality

$$\nabla'^{\mathrm{CKZ}}_{\nu} = f^{\nu} \circ \nabla^{\mathrm{CKZ}}_{\nu} \circ f^{-\nu}$$

holds since

$$\mathrm{d}\, f^{-\nu} = -\nu \sum_{\alpha\in\Delta_+} \mathrm{id}_V \frac{d\alpha}{\alpha}.$$

The connection $\nabla^{\mathrm{CKZ}}_{\nu}$ can be naturally extended on $\widetilde{\mathcal{M}_Q} = \widetilde{\mathfrak{h}_{\mathrm{reg}}}$. As the special case, we consider the CKZ connection on the tangent bundle of $\widetilde{\mathcal{M}_Q}$. We note that the tangent bundle of M is given by

$$T\widetilde{\mathcal{M}_Q} \cong \mathfrak{h} \times \widetilde{\mathcal{M}_Q}.$$

Since the Weyl group W acts on $\mathfrak{h}$ as the reflection group, we have the CKZ connection on $T\widetilde{\mathcal{M}_Q}$.

PROPOSITION 11.8. *The CKZ connection $\nabla^{\mathrm{CKZ}}_{\nu}$ on $T\widetilde{\mathcal{M}_Q}$ coincides with the Dubrovin connection $\check{\nabla}$. In particular, $\nabla^{\mathrm{CKZ}}_{\nu}$ on $T\widetilde{\mathcal{M}_Q}$ is torsion-free and the twisted period P_ν satisfies*

$$\nabla^{\mathrm{CKZ}}_{\nu}\, \mathrm{d}\, P_\nu = 0.$$

PROOF. First we note that under the identification $T\widetilde{\mathcal{M}_Q} \cong \mathfrak{h} \times \widetilde{\mathcal{M}_Q}$, the vector field $\partial_i = \frac{\partial}{\partial p^i}$ is identified with e_i. By using equation (11.4) we have

$$\check{\nabla}_{\partial_i}\partial_j = \nu \partial_i * \partial_j = \nu \sum_{\alpha\in\Delta_+}\sum_{k,l} \frac{\alpha(e_i)\alpha(e_j)\alpha(e_k)}{\alpha(p)} G^{kl}\partial_l.$$

On the other hand,

$$\nabla^{\mathrm{CKZ}}_{\nu,e_i} e_j = -\nu \sum_{\alpha\in\Delta_+} (r_\alpha(e_j) - e_j)\frac{\langle d\alpha, e_i\rangle}{\alpha(p)} = \nu \sum_{\alpha\in\Delta_+} (e_j - r_\alpha(e_j))\frac{\alpha(e_i)}{\alpha(p)}.$$

Since

$$e_j - r_\alpha(e_j) = e_j - (e_j - \alpha(e_j)\alpha^\vee) = \alpha(e_j)\sum_{k,l} \alpha(e_k) G^{kl} e_l,$$

the result follows. □

Thus the twisted periods for Frobenius manifolds of type ADE can be characterized as the flat coordinates of the CKZ connections. Together with Theorem 11.6, we have the following.

COROLLARY 11.9. *The monodromy representation of* Br_Q *for the twisted period map P_ν factors through the Hecke algebra H_Q.*

REMARK 11.10. In [**Sa1**], Saito defined the period map associated with the primitive form on the unfolding of an isolated singularity. He also defined the holonomic D-module (with the complex parameter $s \in \mathbb{C}$) whose solutions characterize his period map. The above twisted period maps P_ν on almost Frobenius manifolds of type ADE correspond to Saito's period maps in the case that isolated singularities are of type ADE. For the details, we refer to [**Sa1**, § 5], [**Sa2**, § 4] and [**D1**, § 5.1].

11.5. Cotangent bundles of the spaces of stability conditions

Let $\mathcal{D}_{\mathbb{X}}$ be a triangulated category satisfying Assumption 2.4 and take a basis $[E_1], \ldots, [E_n]$ of $K(\mathcal{D}_{\mathbb{X}})$. Then

$$K(\mathcal{D}_{\mathbb{X}}) \cong \bigoplus_{i=1}^{n} R[E_i].$$

Consider holomorphic functions

$$Z^i \colon \operatorname{QStab}_s \mathcal{D}_{\mathbb{X}} \to \mathbb{C} \quad (i = 1, \ldots, n)$$

given as the components of the cental charge defined by $Z^i(\sigma) := Z(E_i)$ for $\sigma = (Z, \mathcal{P}) \in \operatorname{QStab}_s \mathcal{D}_{\mathbb{X}}$. Consider holomorphic 1-forms $\mathrm{d}\, Z^1, \ldots, \mathrm{d}\, Z^n$ on $\operatorname{QStab}_s \mathcal{D}_{\mathbb{X}}$ and define a vector space

$$V_{\mathrm{S}} = \bigoplus_{i=1}^{n} \mathbb{C}\, \mathrm{d}\, Z^i.$$

The following is a direct consequence of Theorem 2.5.

PROPOSITION 11.11. *The cotangent bundle of the space* $\operatorname{QStab}_s \mathcal{D}_{\mathbb{X}}$ *is the trivial bundle given by*

$$T^* \operatorname{QStab}_s \mathcal{D}_{\mathbb{X}} \cong V_{\mathrm{S}} \times \operatorname{QStab}_s \mathcal{D}_{\mathbb{X}}.$$

PROOF. By Theorem 2.5, for any small open subset $U \subset \operatorname{QStab}_s \mathcal{D}_{\mathbb{X}}$, functions $Z^1, \ldots, Z^n$ form local coordinates on U. This implies the result. □

If we consider the trivial connection d on $V_{\mathrm{S}} \times \operatorname{QStab}_s \mathcal{D}_{\mathbb{X}}$, the central charges Z^i satisfies the trivial equation

$$\mathrm{d}\,\mathrm{d}\, Z^i = 0.$$

Together with the content of Section 11.2, this suggests the following fundamental problem on the spaces of stability conditions.

PROBLEM 11.12. To find a simply connected complex manifold M and a torsion-free flat connection ∇ such that there is an isomorphism of complex manifolds

$$\phi \colon M \to \operatorname{QStab}_s \mathcal{D}_{\mathbb{X}}$$

satisfying $\phi^* \mathrm{d} = \nabla$.

In the next subsection, we shall give a conjectural answer for $\mathcal{D}_{\mathbb{X}}(Q)$ with type ADE-quiver Q.

If we find an isomorphism ϕ in the above problem, then the flat coordinates P^i of ∇ on M can be identified with the central charges Z^i up to affine transformation. In Section 11.1, we fix the ambiguity of constant shift of twisted periods by the equation (11.2) via the Euler field. Correspondingly, we introduce the Euler field in the side of stability conditions as follows. Let $\frac{\partial}{\partial Z^i}$ be the dual basis of dZ^i. Then the tangent bundle of $\operatorname{QStab}_s \mathcal{D}_{\mathbb{X}}$ is given by

$$T\operatorname{QStab}_s \mathcal{D}_{\mathbb{X}} \cong \left(\bigoplus_{i=1}^{n} \mathbb{C}\frac{\partial}{\partial Z^i} \right) \times \operatorname{QStab}_s \mathcal{D}_{\mathbb{X}}.$$

The Euler vector field $\mathcal{E}$ of charge d on $\operatorname{QStab}_s \mathcal{D}_{\mathbb{X}}$ is defined by

$$\mathcal{E} := \left(\frac{1-d}{2} + \frac{s-2}{2} \right) \sum_{i=1}^{n} Z^i \frac{\partial}{\partial Z^i}.$$

Then central charges Z^i clearly satisfy the quasi-homogeneity condition

$$\mathcal{E} Z^i = \left(\frac{1-d}{2} + \frac{s-2}{2} \right) Z^i.$$

11.6. Conjectural descriptions

Similar to Definition 5.3, we have the Calabi-Yau-$\mathbb{X}$ version of a path algebra.

DEFINITION 11.13. [**IQ**, § 2.4] Let Q be a quiver of type ADE. Denote by $\Gamma_{\mathbb{X}} Q$ the associated Calabi-Yau-$\mathbb{X}$ Ginzburg $\mathbb{Z}^2$-graded differential algebra, $\mathcal{D}_{\mathbb{X}}(Q)$ its finite dimensional derived category and $\operatorname{QStab}_s \mathcal{D}_{\mathbb{X}}(Q)$ the space of q-stability conditions on $\mathcal{D}_{\mathbb{X}}(Q)$.

CONJECTURE 11.14. *Assume that* $\operatorname{Re}(s) \geq 2$ *and set* $\nu = (s-2)/2$. *There is a* Br_Q*-equivariant isomorphism* $\lozenge_s$ *of complex manifolds such that the diagram*

(11.5)
$$\begin{array}{ccc} \widetilde{\mathcal{M}}_Q & \xrightarrow{\ \lozenge_s\ } & \operatorname{QStab}_s^{\circ} \mathcal{D}_{\mathbb{X}}(Q) \\ & {\scriptstyle P_\nu} \searrow \quad \swarrow {\scriptstyle \mathcal{Z}_s} & \\ & \mathbb{C}^n & \end{array}$$

commutes. In particular, $\lozenge_s$ *satisfies that*

$$\begin{cases} \lozenge_s^* \mathcal{E} = E, \\ \lozenge_s^* \mathrm{d} = \nabla_\nu^{\mathrm{CKZ}} \end{cases}$$

for Euler fields and torsion-free flat connections on each side.

11.7. Type A case

When Q is of type A, i.e., $\mathbf{S} = \mathbf{D}$ is a (graded) disk with $n+1$ marked points on its boundary as in Section 10.6. Take a full formal arc system $\mathbf{A}$ of $\mathbf{D}$ and we identify $\mathcal{D}_{\mathbb{X}}(Q) = \mathcal{D}_{\mathbb{X}}(\mathbf{A})$.

Then putting the results of Corollary 10.12 and Proposition 10.15 together, we have the identification

(11.6)
$$\operatorname{QStab}_s^{\circ} \mathcal{D}_{\mathbb{X}}(A_n) \cong \widetilde{\mathcal{M}}_{A_n} = \widetilde{\mathfrak{h}_{\mathrm{reg}}/W}$$

via moduli spaces of (framed) q-quadratic differentials and (framed) Hurwitz spaces. Thus, this implies the following.

THEOREM 11.15. *Assume that* $\operatorname{Re}(s) > 2$. *Let* Q *be the* A_n*-quiver. Then Conjecture* 11.14 *holds.*

REMARK 11.16. For integer valued s, there are following previous works:

- $s = 2$ is essentially shown in [**B5**];
- $s = 3$ is essentially shown in [**BS**];
- $s \in \mathbb{Z}_{\geq 2}$ is shown in [**I**] based on the theory of [**BS**].

Bibliography

[A] C. Amiot, *Cluster categories for algebras of global dimension 2 and quivers with potential* (English, with English and French summaries), Ann. Inst. Fourier (Grenoble) **59** (2009), no. 6, 2525–2590, DOI 10.5802/aif.2499. MR2640929

[ABC+] P. S. Aspinwall, T. Bridgeland, A. Craw, M. R. Douglas, M. Gross, A. Kapustin, G. W. Moore, G. Segal, B. Szendrői, and P. M. H. Wilson, *Dirichlet branes and mirror symmetry*, Clay Mathematics Monographs, vol. 4, American Mathematical Society, Providence, RI; Clay Mathematics Institute, Cambridge, MA, 2009. MR2567952

[AT] P. N. Achar and D. Treumann, *Baric structures on triangulated categories and coherent sheaves*, Int. Math. Res. Not. IMRN **16** (2011), 3688–3743, DOI 10.1093/imrn/rnq226. MR2824842

[BM] A. Bayer and E. Macrì, *The space of stability conditions on the local projective plane*, Duke Math. J. **160** (2011), no. 2, 263–322, DOI 10.1215/00127094-1444249. MR2852118

[BB] A. Bondal and M. van den Bergh, *Generators and representability of functors in commutative and noncommutative geometry* (English, with English and Russian summaries), Mosc. Math. J. **3** (2003), no. 1, 1–36, 258, DOI 10.17323/1609-4514-2003-3-1-1-36. MR1996800

[BrSa] E. Brieskorn and K. Saito, *Artin-Gruppen und Coxeter-Gruppen* (German), Invent. Math. **17** (1972), 245–271, DOI 10.1007/BF01406235. MR0323910

[B1] T. Bridgeland, *Stability conditions on a non-compact Calabi-Yau threefold*, Comm. Math. Phys. **266** (2006), no. 3, 715–733, DOI 10.1007/s00220-006-0048-7. MR2238896

[B2] T. Bridgeland, *Stability conditions on triangulated categories*, Ann. of Math. (2) **166** (2007), no. 2, 317–345, DOI 10.4007/annals.2007.166.317. MR2373143

[B3] T. Bridgeland, *Stability conditions on $K3$ surfaces*, Duke Math. J. **141** (2008), no. 2, 241–291, DOI 10.1215/S0012-7094-08-14122-5. MR2376815

[B4] T. Bridgeland, *Spaces of stability conditions*, Algebraic geometry—Seattle 2005. Part 1, Proc. Sympos. Pure Math., vol. 80, Part 1, Amer. Math. Soc., Providence, RI, 2009, pp. 1–21, DOI 10.1090/pspum/080.1/2483930. MR2483930

[B5] T. Bridgeland, *Stability conditions and Kleinian singularities*, Int. Math. Res. Not. IMRN **21** (2009), 4142–4157, DOI 10.1093/imrn/rnp081. MR2549952

[BQS] T. Bridgeland, Y. Qiu, and T. Sutherland, *Stability conditions and the A_2 quiver*, Adv. Math. **365** (2020), 107049, 33, DOI 10.1016/j.aim.2020.107049. MR4064773

[BS] T. Bridgeland and I. Smith, *Quadratic differentials as stability conditions*, Publ. Math. Inst. Hautes Études Sci. **121** (2015), 155–278, DOI 10.1007/s10240-014-0066-5. MR3349833

[BZ] T. Brüstle and J. Zhang, *On the cluster category of a marked surface without punctures*, Algebra Number Theory **5** (2011), no. 4, 529–566, DOI 10.2140/ant.2011.5.529. MR2870100

[BMR+] A. B. Buan, R. Marsh, M. Reineke, I. Reiten, and G. Todorov, *Tilting theory and cluster combinatorics*, Adv. Math. **204** (2006), no. 2, 572–618, DOI 10.1016/j.aim.2005.06.003. MR2249625

[BQZ] A. B. Buan, Y. Qiu, and Y. Zhou, *Decorated marked surfaces III: The derived category of a decorated marked surface*, Int. Math. Res. Not. IMRN **17** (2021), 12967–12992, DOI 10.1093/imrn/rnz180. MR4307679

[CL] G. Cerulli Irelli and D. Labardini-Fragoso, *Quivers with potentials associated to triangulated surfaces, Part III: tagged triangulations and cluster monomials*, Compos. Math. **148** (2012), no. 6, 1833–1866, DOI 10.1112/S0010437X12000528. MR2999307

[C] I. V. Cherednik, *Generalized braid groups and local r-matrix systems* (Russian), Dokl. Akad. Nauk SSSR **307** (1989), no. 1, 49–53; English transl., Soviet Math. Dokl. **40** (1990), no. 1, 43–48. MR1017085

[DWZ] H. Derksen, J. Weyman, and A. Zelevinsky, *Quivers with potentials and their representations. I. Mutations*, Selecta Math. (N.S.) **14** (2008), no. 1, 59–119, DOI 10.1007/s00029-008-0057-9. MR2480710

[DHKK] G. Dimitrov, F. Haiden, L. Katzarkov, and M. Kontsevich, *Dynamical systems and categories*, The influence of Solomon Lefschetz in geometry and topology, Contemp. Math., vol. 621, Amer. Math. Soc., Providence, RI, 2014, pp. 133–170, DOI 10.1090/conm/621/12421. MR3289326

[D1] B. Dubrovin, *Geometry of 2D topological field theories*, Integrable systems and quantum groups (Montecatini Terme, 1993), Lecture Notes in Math., vol. 1620, Springer, Berlin, 1996, pp. 120–348, DOI 10.1007/BFb0094793. MR1397274

[D2] B. Dubrovin, *On almost duality for Frobenius manifolds*, Geometry, topology, and mathematical physics, Amer. Math. Soc. Transl. Ser. 2, vol. 212, Amer. Math. Soc., Providence, RI, 2004, pp. 75–132, DOI 10.1090/trans2/212/05. MR2070050

[FM] D. Fiorenza and G. Marchetti, *The (he)art of gluing* Preprint, `arXiv:1806.00883`, (2018).

[FST] S. Fomin, M. Shapiro, and D. Thurston, *Cluster algebras and triangulated surfaces. I. Cluster complexes*, Acta Math. **201** (2008), no. 1, 83–146, DOI 10.1007/s11511-008-0030-7. MR2448067

[GMN] D. Gaiotto, G. W. Moore, and A. Neitzke, *Wall-crossing, Hitchin systems, and the WKB approximation*, Adv. Math. **234** (2013), 239–403, DOI 10.1016/j.aim.2012.09.027. MR3003931

[Gu] L. Guo, *Cluster tilting objects in generalized higher cluster categories*, J. Pure Appl. Algebra **215** (2011), no. 9, 2055–2071, DOI 10.1016/j.jpaa.2010.11.015. MR2786597

[H] F. Haiden, Personal communication (emails).

[HKK] F. Haiden, L. Katzarkov, and M. Kontsevich, *Flat surfaces and stability structures*, Publ. Math. Inst. Hautes Études Sci. **126** (2017), 247–318, DOI 10.1007/s10240-017-0095-y. MR3735868

[HK] Z. Hua and B. Keller, *Cluster categories and rational curves*, Geom. Topol. **28** (2024), no. 6, 2569–2634, DOI 10.2140/gt.2024.28.2569. MR4817470

[I] A. Ikeda, *Stability conditions on* CY_N *categories associated to* A_n*-quivers and period maps*, Math. Ann. **367** (2017), no. 1-2, 1–49, DOI 10.1007/s00208-016-1375-4. MR3606433

[IQ] A. Ikeda and Y. Qiu, *q-stability conditions on Calabi-Yau-*$\mathbb{X}$ *categories*, Compos. Math. **159** (2023), no. 7, 1347–1386, DOI 10.1112/s0010437x23007194. MR4599210

[IQZ] A. Ikeda, Y. Qiu, and Y. Zhou, *Graded decorated marked surfaces: Calabi-Yau-*$\mathbb{X}$ *categories of gentle algebras*, Preprint, `arXiv:2006.00009`, (2020).

[IUU] A. Ishii, K. Ueda, and H. Uehara, *Stability conditions on* A_n*-singularities*, J. Differential Geom. **84** (2010), no. 1, 87–126. MR2629510

[IY] O. Iyama and D. Yang, *Quotients of triangulated categories and equivalences of Buchweitz, Orlov, and Amiot-Guo-Keller*, Amer. J. Math. **142** (2020), no. 5, 1641–1659, DOI 10.1353/ajm.2020.0041. MR4150654

[LP] Y. Lekili and A. Polishchuk, *Auslander orders over nodal stacky curves and partially wrapped Fukaya categories*, J. Topol. **11** (2018), no. 3, 615–644, DOI 10.1112/topo.12064. MR3830878

[LP2] Y. Lekili and A. Polishchuk, *Derived equivalences of gentle algebras via Fukaya categories*, Math. Ann. **376** (2020), no. 1-2, 187–225, DOI 10.1007/s00208-019-01894-5. MR4055159

[KaY] M. Kalck and D. Yang, *Derived categories of graded gentle one-cycle algebras*, J. Pure Appl. Algebra **222** (2018), no. 10, 3005–3035, DOI 10.1016/j.jpaa.2017.11.011. MR3795632

[K1] B. Keller, *Deformed Calabi-Yau completions*, J. Reine Angew. Math. **654** (2011), 125–180, DOI 10.1515/CRELLE.2011.031. With an appendix by Michel Van den Bergh. MR2795754

[K2] B. Keller, *On triangulated orbit categories*, Doc. Math. **10** (2005), 551–581. MR2184464

[K3] B. Keller, *On cluster theory and quantum dilogarithm identities*, Representations of algebras and related topics, EMS Ser. Congr. Rep., Eur. Math. Soc., Zürich, 2011, pp. 85–116, DOI 10.4171/101-1/3. MR2931896

[K4] B. Keller, *Erratum to 'Deformed Calabi-Yau completions'*, `arXiv:1109.2924`.

[K5] B. Keller, *On differential graded categories*, International Congress of Mathematicians. Vol. II, Eur. Math. Soc., Zürich, 2006, pp. 151–190. MR2275593

[K6] B. Keller, *Calabi-Yau triangulated categories*, Trends in representation theory of algebras and related topics, EMS Ser. Congr. Rep., Eur. Math. Soc., Zürich, 2008, pp. 467–489, DOI 10.4171/062-1/11. MR2484733

[KQ1] A. King and Y. Qiu, *Exchange graphs and Ext quivers*, Adv. Math. **285** (2015), 1106–1154, DOI 10.1016/j.aim.2015.08.017. MR3406522

[KQ2] A. King and Y. Qiu, *Cluster exchange groupoids and framed quadratic differentials*, Invent. Math. **220** (2020), no. 2, 479–523, DOI 10.1007/s00222-019-00932-y. MR4081136

[KhS] M. Khovanov and P. Seidel, *Quivers, Floer cohomology, and braid group actions*, J. Amer. Math. Soc. **15** (2002), no. 1, 203–271, DOI 10.1090/S0894-0347-01-00374-5. MR1862802

[Ko] M. Kontsevich, *Homological algebra of mirror symmetry*, Proceedings of the International Congress of Mathematicians, Vol. 1, 2 (Zürich, 1994), Birkhäuser, Basel, 1995, pp. 120–139. MR1403918

[KoS] M. Kontsevich and Y. Soibelman, *Stability structures, motivic Donaldson-Thomas invariants and cluster transformations*, Preprint, `arXiv:0811.2435`, (2008).

[M] Y. I. Manin, *Frobenius manifolds, quantum cohomology, and moduli spaces*, American Mathematical Society Colloquium Publications, vol. 47, American Mathematical Society, Providence, RI, 1999, DOI 10.1090/coll/047. MR1702284

[Op] S. Oppermann, *Quivers for silting mutation*, Adv. Math. **307** (2017), 684–714, DOI 10.1016/j.aim.2016.11.024. MR3590528

[Or] D. Orlov, *Remarks on generators and dimensions of triangulated categories* (English, with English and Russian summaries), Mosc. Math. J. **9** (2009), no. 1, 153–159, back matter, DOI 10.17323/1609-4514-2009-9-1-143-149. MR2567400

[Q1] Y. Qiu, *Stability conditions and quantum dilogarithm identities for Dynkin quivers*, Adv. Math. **269** (2015), 220–264, DOI 10.1016/j.aim.2014.10.014. MR3281136

[Q2] Y. Qiu, *Decorated marked surfaces: spherical twists versus braid twists*, Math. Ann. **365** (2016), no. 1-2, 595–633, DOI 10.1007/s00208-015-1339-0. MR3498923

[Q3] Y. Qiu, *Decorated marked surfaces (part B): topological realizations*, Math. Z. **288** (2018), no. 1-2, 39–53, DOI 10.1007/s00209-017-1876-1. MR3774402

[Q4] Y. Qiu, *Topological structure of spaces of stability conditions and topological Fukaya type categories*, Proceedings of the International Consortium of Chinese Mathematicians 2017, Int. Press, Boston, MA, [2020] ©2020, pp. 521–538. MR4251127

[Q5] Y. Qiu, *Global dimension function on stability conditions and Gepner equations*, Math. Z. **303** (2023), no. 1, Paper No. 11, 24, DOI 10.1007/s00209-022-03170-w. MR4519568

[Q6] Y. Qiu, *Contractible flow of stability conditions via global dimension function*, J. Differential Geom. **129** (2025), no. 2, 491–521, DOI 10.4310/jdg/1738163897. MR4859581

[QW] Y. Qiu and J. Woolf, *Contractible stability spaces and faithful braid group actions*, Geom. Topol. **22** (2018), no. 6, 3701–3760, DOI 10.2140/gt.2018.22.3701. MR3858773

[QZ1] Y. Qiu and Y. Zhou, *Cluster categories for marked surfaces: punctured case*, Compos. Math. **153** (2017), no. 9, 1779–1819, DOI 10.1112/S0010437X17007229. MR3705277

[QZ2] Y. Qiu and Y. Zhou, *Decorated marked surfaces II: Intersection numbers and dimensions of Homs*, Trans. Amer. Math. Soc. **372** (2019), no. 1, 635–660, DOI 10.1090/tran/7598. MR3968782

[Sa1] K. Saito, *Period mapping associated to a primitive form*, Publ. Res. Inst. Math. Sci. **19** (1983), no. 3, 1231–1264, DOI 10.2977/prims/1195182028. MR0723468

[Sa2] K. Saito, *Uniformization of the orbifold of a finite reflection group*, Frobenius manifolds, Aspects Math., vol. E36, Friedr. Vieweg, Wiesbaden, 2004, pp. 265–320. MR2115774

[SaTa] K. Saito and A. Takahashi, *From primitive forms to Frobenius manifolds*, From Hodge theory to integrability and TQFT tt*-geometry, Proc. Sympos. Pure Math., vol. 78, Amer. Math. Soc., Providence, RI, 2008, pp. 31–48, DOI 10.1090/pspum/078/2483747. MR2483747

[Se] P. Seidel, *Picard-Lefschetz theory and dilating $\mathbb{C}^*$-actions*, J. Topol. **8** (2015), no. 4, 1167–1201, DOI 10.1112/jtopol/jtv029. MR3431673

[ST] P. Seidel and R. Thomas, *Braid group actions on derived categories of coherent sheaves*, Duke Math. J. **108** (2001), no. 1, 37–108, DOI 10.1215/S0012-7094-01-10812-0. MR1831820

[TL] V. Toledano Laredo, *Flat connections and quantum groups*, Acta Appl. Math. **73** (2002), no. 1-2, 155–173, DOI 10.1023/A:1019791123828. The 2000 Twente Conference on Lie Groups (Enschede). MR1926499

[Ta] A. Takeda, *Relative stability conditions on Fukaya categories of surfaces*, Math. Z. **301** (2022), no. 3, 3019–3070, DOI 10.1007/s00209-022-03007-6. MR4437347

[To] Y. Toda, *Gepner type stability conditions on graded matrix factorizations*, Algebr. Geom. **1** (2014), no. 5, 613–665, DOI 10.14231/AG-2014-026. MR3296807

[S] I. Smith, *Quiver algebras as Fukaya categories*, Geom. Topol. **19** (2015), no. 5, 2557–2617, DOI 10.2140/gt.2015.19.2557. MR3416110

[Y] W-K. Yeung, *Relative Calabi-Yau completions*, Preprint, `arXiv:1612.06352`, (2022).

Editorial Information

To be published in the *Memoirs*, a paper must be correct, new, nontrivial, and significant. Further, it must be well written and of interest to a substantial number of mathematicians. Piecemeal results, such as an inconclusive step toward an unproved major theorem or a minor variation on a known result, are in general not acceptable for publication.

Papers appearing in *Memoirs* are generally at least 80 and not more than 200 published pages in length. Papers less than 80 or more than 200 published pages require the approval of the Managing Editor of the Transactions/Memoirs Editorial Board. Published pages are the same size as those generated in the style files provided for $\mathcal{A}\mathcal{M}\mathcal{S}$-LaTeX.

Information on the backlog for this journal can be found on the AMS website starting from `http://www.ams.org/memo`.

A Consent to Publish is required before we can begin processing your paper. After a paper is accepted for publication, the Providence office will send a Consent to Publish and Copyright Agreement to all authors of the paper. By submitting a paper to the *Memoirs*, authors certify that the results have not been submitted to nor are they under consideration for publication by another journal, conference proceedings, or similar publication.

Information for Authors

Memoirs is an author-prepared publication. Once formatted for print and on-line publication, articles will be published as is with the addition of AMS-prepared frontmatter and backmatter. Articles are not copyedited; however, confirmation copy will be sent to the authors.

Initial submission. The AMS uses Centralized Manuscript Processing for initial submissions. Authors should submit a PDF file using the Initial Manuscript Submission form found at `www.ams.org/submission/memo`, or send one copy of the manuscript to the following address: Centralized Manuscript Processing, MEMOIRS OF THE AMS, 201 Charles Street, Providence, RI 02904-2294 USA. If a paper copy is being forwarded to the AMS, indicate that it is for *Memoirs* and include the name of the corresponding author, contact information such as email address or mailing address, and the name of an appropriate Editor to review the paper (see the list of Editors below).

The paper must contain a *descriptive title* and an *abstract* that summarizes the article in language suitable for workers in the general field (algebra, analysis, etc.). The *descriptive title* should be short, but informative; useless or vague phrases such as "some remarks about" or "concerning" should be avoided. The *abstract* should be at least one complete sentence, and at most 300 words. Included with the footnotes to the paper should be the 2020 *Mathematics Subject Classification* representing the primary and secondary subjects of the article. The classifications are accessible from `www.ams.org/msc/`. The Mathematics Subject Classification footnote may be followed by a list of *key words and phrases* describing the subject matter of the article and taken from it. Journal abbreviations used in bibliographies are listed in the latest *Mathematical Reviews* annual index. The series abbreviations are also accessible from `www.ams.org/msnhtml/serials.pdf`. To help in preparing and verifying references, the AMS offers MR Lookup, a Reference Tool for Linking, at `www.ams.org/mrlookup/`.

Electronically prepared manuscripts. The AMS encourages electronically prepared manuscripts, with a strong preference for $\mathcal{A}\mathcal{M}\mathcal{S}$-LaTeX. To this end, the Society has prepared $\mathcal{A}\mathcal{M}\mathcal{S}$-LaTeX author packages for each AMS publication. Author packages include instructions for preparing electronic manuscripts, samples, and a style file that generates the particular design specifications of that publication series.

Authors may retrieve an author package for *Memoirs of the AMS* from `www.ams.org/journals/memo/memoauthorpac.html`. The *AMS Author Handbook* is available in PDF format from the author package link. The author package can also be obtained free of charge by sending email to `tech-support@ams.org` or from the Publication Division,

American Mathematical Society, 201 Charles St., Providence, RI 02904-2294, USA. When requesting an author package, please specify the publication in which your paper will appear. Please be sure to include your complete mailing address.

After acceptance. The source files for the final version of the electronic manuscript should be sent to the Providence office immediately after the paper has been accepted for publication. The author should also submit a PDF of the final version of the paper to the editor, who will forward a copy to the Providence office.

Accepted electronically prepared files can be submitted via the web at `www.ams.org/submit-book-journal/`, sent via FTP, or sent on CD to the Electronic Prepress Department, American Mathematical Society, 201 Charles Street, Providence, RI 02904-2294 USA. TEX source files and graphic files can be transferred over the Internet by FTP to the Internet node `ftp.ams.org` (130.44.1.100). When sending a manuscript electronically via CD, please be sure to include a message indicating that the paper is for the *Memoirs*.

Electronic graphics. Comprehensive instructions on preparing graphics are available at `www.ams.org/authors/journals.html`. A few of the major requirements are given here.

Submit files for graphics as EPS (Encapsulated PostScript) files. This includes graphics originated via a graphics application as well as scanned photographs or other computer-generated images. If this is not possible, TIFF files are acceptable as long as they can be opened in Adobe Photoshop or Illustrator.

Authors using graphics packages for the creation of electronic art should also avoid the use of any lines thinner than 0.5 points in width. Many graphics packages allow the user to specify a "hairline" for a very thin line. Hairlines often look acceptable when proofed on a typical laser printer. However, when produced on a high-resolution laser imagesetter, hairlines become nearly invisible and will be lost entirely in the final printing process.

Screens should be set to values between 15% and 85%. Screens which fall outside of this range are too light or too dark to print correctly. Variations of screens within a graphic should be no less than 10%.

Any graphics created in color will be rendered in grayscale for the printed version unless color printing is authorized by the Managing Editor and the Publisher. In general, color graphics will appear in color in the online version.

Inquiries. Any inquiries concerning a paper that has been accepted for publication should be sent to `memo-query@ams.org` or directly to the Electronic Prepress Department, American Mathematical Society, 201 Charles St., Providence, RI 02904-2294 USA.

SELECTED PUBLISHED TITLES IN THIS SERIES

For a complete list of titles in this series, visit the
AMS Bookstore at **www.ams.org/bookstore/memoseries/**.